Gebhard Geiger

Evolutionary Instability

Logical and Material Aspects of a Unified
Theory of Biosocial Evolution

With 20 Figures

Springer-Verlag Berlin Heidelberg New York
London Paris Tokyo Hong Kong

Dr. GEBHARD GEIGER
Hohenzollernplatz 8
D-8000 München 40
Fed. Rep. of Germany

ISBN 3-540-51808-8 Springer-Verlag Berlin Heidelberg New York
ISBN 0-387-51808-8 Springer-Verlag New York Berlin Heidelberg

Library of Congress Cataloging-in-Publication Data. Geiger, Gebhard. Evolutionary instability: logical and material aspects of a unified theory of biosocial evolution / Gebhard Geiger. p. cm. Includes bibliographical references. ISBN 0-387-51808-8 (alk. paper) 1. Human evolution--Philosophy. 2. Social evolution. 3. Sociobiology. I. Title. GN281.4.G45 1990 303.4--dc20

This work is subject to copyright. All rights are reserved, whether the whole or part of the material is concerned, specifically the rigths of translation, reprinting, re-use of illustrations, recitation, broadcasting, reproduction on microfilms or in other ways, and storage in data banks. Duplication of this publication or parts thereof is only permitted under the provisions of the German Copyright Law of September 9, 1965, in its version of June 24, 1985, and a copyright fee must always be paid. Violations fall under the prosecution act of the German Copyright Law.

© Springer-Verlag Berlin Heidelberg 1990
Printed in Germany

The use of registered names, trademarks, etc. in this publication does not imply, even in the absence of a specific statement, that such names are exempt from the relevant protective laws and regulations and therefore free for general use.

Typesetting: Thomson Press (India) Ltd. New Delhi
2131/3145-(3011)543210 — Printed on acid-free paper

Für Gisela

Preface

The scope of this book can be described best as a compilation of papers presented to an imaginary interdisciplinary conference on philosophical and material problems of biosocial evolution. The conference is especially designed to discuss unifying theoretical approaches to phenomena of both increasing structural complexity („natural self-organisation") and biosocial adaptation. Among the contributors to the conference are mathematical system theoretists, philosophers of science, theoretical population biologists, and social scientists. The only difference to a real conference of this kind is that all contributions are delivered by one and the same author, who also tries to integrate them to some higher degree than can normally be attained in conference papers. Technically, this integration amounts to the embedding of the biobehavioural concepts of evolutionarily stable and unstable strategies into the framework of the structure and stability of hierarchical systems.

When embarking on my work on philosophical and interdisciplinary problems of the life sciences about 6 years ago, I decided not to contribute to the obscurantist talk of – *horribile dictu* – „emergence", „ontological reductionism" and the like contaminating the recent sociobiology debate. I rather found it a genuinely philosophical task to *analyse* this talk, and so I attempted to recast the relevant concepts and principles into a form allowing for definite conclusions. Among other things, which I hope the reader will find more pleasant, this attempt had some unfortunate effect, however. It led to considerable formalisations in the system-theoretical and philosophical parts of the analysis. I therefore took additional measures to make the premises, arguments, and results of my analysis intuitively more accessible. Most of the formalised definitions and inferences are restated in non-technical terms within the text, and illustrated by familiar examples and graphical representations. Each section using the formal approach is supplemented by an informal summary and discussion. Chapter VI on the evolution of human social structure, which is entirely kept in informal terms, once more rephrases the central concepts of structural and evolutionary stability and instability in sufficient, non-mathematical detail. I hope that these restaments make each part of the inquiry appear more self-contained rather than excessively repetitious.

Considering its broad interdisciplinary scope, the book provides explicit definitions only of those concepts on whose analysis I wish to concentrate. All other concepts, principles and theories referred to in the text I suppose to be explained elsewhere (e.g., in the literature cited). Any attempt at greater completeness in the exposition of my theory would have blown up the size of the book — and that of its subject index — at least by a factor of three. A similar point can be made concerning the list of references. Any attempt at a complete coverage of the recent literature on biosocial evolution and its exceeding interdisciplinary complexities seemed illusory to me. So I hope I at least succeeded in sampling a number of books and articles that have made major contributions to the recent sociobiology debate.

The urge to treat each aspect of my theme in problem-oriented technical terms also has some less unfortunate implications, however. When considering biological and sociocultural applications of the analytical framework established in Part One, I no longer felt obliged to follow the formal logical approach and adopted a style of writing common among working scientists. For instance, in Part Two definitions tend to be introduced in a less explicit, more intuitive fashion, and the philosophical distinction between object and metalanguage is handled more casually. The style of Part One may be more exact, but that of Part Two is less pedantic and, hopefully, easier to digest.

My work would have been impossible without the support of many friends and colleagues whose help, encouragement and stimulating interest I gratefully acknowledge. I am particularly indebted to Professor Jan Berg, *Institut für Philosophie der Technischen Universität München*, in whose institute I completed my *Habilitationsschrift* in philosophy of science on which Part One of this book is based. His, and Professor Reinhard Kleinknecht's, numerous useful comments on many aspects of my work, discussions and criticisms were of great help in carrying out the philosophical part of my inquiry.

Above all I thank my wife Gisela, to whom my work owes much of the evolutionary stability necessary for book manuscripts to grow into the age, or stage, of reproduction.

GEBHARD GEIGER

Contents

I   **Introduction**   1

Part One   Structure and Evolution of Hierarchical Systems

II   **On Emergent Structures, Truisms and Fallacies**   7

1   Structure and Complexity of Evolving Systems   7
1.1   Examples   7
1.2   Basic Definitions   10
1.3   Informal Summary   24

2   Basic Problems of the Evolution of Matter   25

3   Holism Versus Reductionism   27
3.1   Reduction of Theroies   29
3.2   Lower-Level Representations   32
3.3   Cross-Level Deterministic Effects   33
3.4   Holism Reconsidered   34
3.5   Informal Statement of Results   36

4   The Paradox of Emergent Evolution   37
4.1   Emergent Evolution: A Crude Misconception   37
4.2   A Note on Stochastic Evolutionary Theories   38

III   **The Concept of Unified Theory**   40

1   Synthesis Versus Reduction of Theories   40

2   Synthesis of Theories Through Parametrisation of Laws   42
2.1   Parametrisation of Laws   42
2.2   Synthesis of Theories   43
2.3   Informal Summary   44

3   The Central Representation Theorem   45
3.1   The Meaning and Significance of the Theorem: An Informal Résumé   51

4	State-Determined Hierarchical Systems	52
4.1	Parameter Families of Systems	52
4.2	Unified System Theories	54
4.3	State-Determined Hierarchies	55
4.4	Evolutionary Processes	56
4.5	Informal Summary and Discussion	58
5	Pointless Scientific Controversies	58
5.1	Cross-Level Similarity of Structures	59
5.2	Cultural Systems as "Emergent Wholes"	60
5.3	System Simulation	61
5.4	The Problem of Similarity in Comparative Ethology	63
5.5	Darwinism Versus Mendelism	63

Appendix: Examples 65

1	Interacting Biological Populations	65
1.1	Dynamical Systems	65
1.2	Interacting Populations as Coupled Systems	67
1.3	The Associate Parameter Family	68
1.4	State Determinacy and Hierarchical Evolution	69

Part Two   The Evolution of Social Structure

IV	**Perspectives on Non-Adaptive Evolution**	73
1	Is Sociobiology Reductionist?	73
1.1	Complexities of Social Interaction	73
1.2	Sociobiology: Merits and Limits	74
1.3	The Quest for Alternatives	75
1.4	An Approach to Non-Adaptive Change	78
2	The Concepts of Structural and Evolutionary Instability	78
2.1	The Meaning of Structural Instability	79
2.2	Structurally Stable and Unstable Games	80
2.3	Informal Summary and Discussion	82
V	**Structural Instability in Evolutionary Population Biology**	84
1	Sociobiology and the Structural Instability of Behaviour Patterns	84
1.1	Sources of Evolutionary Change	84
1.2	A Dynamical Approach to Biosocial Genetics	86
1.3	Asymptotically Stable Equilibria	89
1.4	Favourable Mutations	90

1.5	Application to Inclusive-Fitness Theory	93
1.6	Applications to Insect Social Structure	95
1.7	Evolutionary Instability in Secular Time Scales	97
1.8	Informal Summary and Conclusion	101
2	Structural Instability in Population Dynamics	102
2.1	Population Interactions in Secular Time Scales	104
2.2	Low-Dimensional Examples	108
2.3	The Adaptive Topography Reconstructed	114
2.4	Informal Summary and Discussion	116
3	Dynamics and Structural Change in Biocultural Coevolution	117
3.1	The Basic Equations	119
3.2	The Concept of Cultural Capacity	122
3.3	The Impact of Learning on Biocultural Evolution	125
3.4	The Expiration of the Coevolutionary Circuit	127
3.5	Summary and Conclusions	129
VI	**Applications to Human Social Structure**	**131**
1	The Anthropological Significance of Evolutionary Stability and Instability	131
2	Political Power as an Evolutionary Structure	133
2.1	Evolution of Hierarchical Complexity	135
2.2	Stability of Social Structure	137
2.3	Biobehavioural Bases of Influence and Power	140
2.4	Political Power	143
2.5	Summary and Discussion	146
3	On the Evolution of Complex, Political Society	147
3.1	The Evolutionary Conceptual Framework	147
3.2	Biobehavioural Bases of Sociocultural Complexity	149
3.3	From Rites to Sanctions	150
3.4	Summary	151
VII	**Concluding Remarks**	**152**

References 155

List of Symbols 163

Subject Index 165

I Introduction

The notion that living systems, including human individuals, cultures and societies, are the outcome of the physical properties of matter, is at least as old as empirical science itself. To be sure, in the history of science and philosophy one rarely finds mechanistic, or physicalistic, views of the organic world which, in simple-minded manners, held no more than the machine-like nature of organisms. These views were rather based on elaborate epistemologies and systematics of science, with classical physics and mechanics conceived of as being paradigmatic of all human inquiry. In any case, efforts to systematise science and epistemology on the basis of the methods and laws of physics do not only characterise modern scientific and philosophical positivism and materialism. Attempts in this direction were already carried out at great length and sophistication in Thomas Hobbes' *Elementa Philosophiae* and *Leviathan*, to recall to just one of the classical examples.

Although classical mechanics has now lost much of its paradigmatic role in empirical science, certain controversial issues inherent in the mechanistic conceptions of the world continue governing scientific and popular debates on the structure of matter and the principles of life. This situation is largely due to the influence of 20th century evolutionism on the kind of research actually carried out in the natural sciences during the recent decades. Roughly speaking, "evolutionism" means belief in the universal applicability of concepts and theories of evolution. In virtually all of the relevant disciplines, the old mechanistic approaches to the structure of matter and life have now been replaced by the search for evolutionary explanations. And for all that we know, the current hypotheses on prebiotic chemical evolution in cosmological time, biomolecular and organic evolution in geological time, and sociocultural evolution in historical time belong to the most successful and empirically satisfying theories ever invented by scientists.

Yet there are scientists and philosophers who blame modern evolutionary thought for misconceptions similar to those inherent in the old physicalistic and mechanistic views of nature. Such blames typically aim at biological, evolutionary accounts of human mental, cultural and social structures. But in a broader sense, they are directed against the so-called reductionist explanations of the hierarchical organisation of matter. The concept of hierarchical organisation of matter intends the distinction between microscopic and macroscopic structures in physical and organic systems, while the term "reductionist" essentially refers to microphysical explanations of the macroscopic attributes of matter. The relevant concepts will be explicitly defined and thoroughly analysed in the following chapters. In a rough classification, two theoretical positions can be distinguished in the contemporary debate on the origins and modes of the hierarchical self-organisation of nature. One position conveys elements of reductionism with regard to the proposed, or one's preferred, explanations of the structure and evolution of hierarchical systems. Examples are the genetic and physiological approaches to human behaviour which have recently stimulated considerable controversy (human ethology, behavioural physiology and sociobiology). The alternative position is based on the postulate that the physical attributes of the constituent parts (microstructure) of a system do not completely determine this system's macroscopic structure and behaviour. This position has also been subsumed under the more traditional concept of holism, corresponding to the Aristotelian notion that the whole is more than the sum of its parts ($\mathring{o}\lambda ov$ = the whole).

The "pretensiousness" of the evolutionary approach in the natural and social sciences has met with all sorts of criticism, ranging from scientific to religious to politically motivated objections. The starting point of the present investigation is the critique of these objections insofar as they conform to the methodological standards of empirical science. Since recent developments in general system theory have strongly reinforced the holistic views of the hierarchical structure of matter (in fact, systems science is sometimes understood as the "general science of wholeness"; von Bertalanffy 1968, p. 37), system-theoretic, holistic contributions to evolutionary debates are given special attention. The mathematical representations of hierarchical structures and systems available in the literature are reorganised in a systematic manner and, where necessary, extended, made more precise, and adapted to the purposes of the present inquiry. The system-theoretic framework is then employed in the logical and semantic analysis of various holistic modes of explanation in the physical and life sciences.

Since in the following chapters the semantics of evolutionary theories receives largely formal treatment, the use of model-theoretic concepts and techniques suggests itself. As customary in logic and mathematics, by model theory we mean that part of the semantics of formalised languages and theories which is concerned with the relations between sentences and mathematical structures in which these sentences are valid. For the sake of definiteness, most of the model-theoretic concepts and theorems referred to are supposed to be well known, and will be made use of without much fuss. The relevant definitions and results belong to the subject matter of most textbooks of elementary logic and model theory. Here we largely draw upon Schwabhäuser's (1971, 1972) introductory texts.

Section II.1 lays down the system-theoretic framework on which the entire investigation is built. Special emphasis is given to the explicit definition and analysis of the concept of hierarchical organisation, comparative concepts of complexity of (hierarchical) structures and systems, and coupling relations between input-output systems. The concepts of decoupled and coupled systems are used to explicate the notoriously cumbersome notions of part and whole. These concepts are related to the hierarchical structure and evolution of prebiotic and living systems in Section II.2. In the next two sections of Chapter II, various holistic connotations of "evolution" are analysed in terms of the system-theoretic framework developed thus far.

Chapter III is concerned with the operations called "synthesis of theories" and "parametrisation of laws" designed to match the logical and semantic structures of different theories. It is shown that syntheses of theories, or unified theories, are particularly suitable to describe hierarchically organised structures and systems. Section III.1 introduces unified theories in general model-theoretic terms. The subclass of unified theories characterised in Section III.2 is significant of theoretical advance in a variety of empirical disciplines. Selected problems, examples and applications are treated in two further sections of Chapter III and, at some more length, in the Appendix. The applications comprise parameter representations of systems, hierarchies and evolutionary processes.

The main result with respect to the holism-reductionism controversy in evolutionary science is expressed by the Central Representation Theorem (Sect. III.3). The theorem states, firstly, that the macrostructures of hierarchical systems can always be given microscopic characterisations of such a kind that the global structure and behaviour of any such system become represented as attributes of the system's constituent parts. The proof of this part of the theorem is constructive in the sense that it makes the intended microrepresentations explicit. The second part of the theorem refers to deterministic theories with first-order logical syntax (i.e., elementary theories) which are semantically interpreted relative to hierarchically ordered structures. Roughly speaking, the theorem asserts that theories of macrostructures can be transformed into theories of microstructures without loss in material content.

The restriction of the second part of the theorem to deterministic theories does not impair the significance of the theorem for empirical theories of hierarchical organisation. Many of the currently investigated questions of the structure and evolution of matter indeed require elaborate stochastic approaches whose logic and semantics extend far beyond the scope of the present inquiry. But it is also well known that approximate deterministic descriptions are available for many critical evolutionary phenomena in biophysics, molecular and population genetics, etc., so that the following chapters by no means lead besides the point. In fact, the examples and applications presented in the Appendix (Part One) and in Part Two are drawn exclusively from the theory of deterministic dynamical systems.

The other results of Part One must be understood in the context of the Representation Theorem. Whether the macroscopic properties of matter possess, or can be given, microphysical representations or not, does not depend so much on the structure of matter itself, but on the theories available to describe it. Accordingly, reducibility is strictly treated as a relation between theories under their respective semantic interpretations rather than between structures. In this respect, we comply with Ernest Nagel's (1979, p. 363) plea for a "linguistic turn" in evolutionary debates.

Similarly, distinctions between microscopic and macroscopic features of hierarchical systems cannot be drawn in an absolute sense, as the opponents to the hypothesis of natural hierarchical self-organisation insinuate. For in a trivial sense one can always define functions which map one structure into another one, that is, represent one structure in terms of the other. Moreover, one may restrict the real-world structures of one's (theoretical) interest so that sufficiently simple representing functions, viz. approximate representations, can be construed (e.g., mathematically). Thus, as a matter of principle, microrepresentations of macrostructures always exist, and, as the Representation Theorem shows, structures confer this property on theories and their semantic interpretations. Strictly speaking, the concept of hierarchical organisation of matter refers to the fact that we may assume microrepresentations or macrorepresentations of the physical world in the empirical interpretations of our scientific theories. Which kind of representation scientists actually choose is a matter of expedience and similar pragmatic criteria rather than a question of whatever absolute order of nature.

On the other hand, the Central Representation Theorem and the associate results derived below do not automatically support reductionist views of natural self-organisation. This situation is an immediate outcome of the present limitation of the concept of reduction to intertheory relations. The transformations of macrophysical theories into microphysical ones carried out in proof of the Representation Theorem, generally do not constitute relationships of reducibility in the sense of model theory.

Our results imply, however, that the holistic views of the hierarchical structure of matter are incompatible with any attempt to account, in evolutionary terms, for this organisational characteristic of nature.

While Part One suggests theoretical alternatives to the doctrines of holism and reductionism, Part Two elaborates upon such alternatives, with special emphasis on problems and applications selected from the ongoing sociobiology debate. It will be demonstrated that the conceptual framework developed in Part One is suitable to tackle some of the most controversial questions of a unified theory of biosocial evolution. Section IV.2 connects the notion of system evolution (Part One) with the biobehavioural conceptual framework of evolutionary stability and instability ("evolutionary game theory"). In the remaining chapters of Part Two, the results of Section IV.2 are extended and applied to various problems of biosocial and biocultural evolution.

Part One
Structure and Evolution of Hierarchical Systems

II On Emergent Structures, Truisms and Fallacies

1 Structure and Complexity of Evolving Systems

Throughout the present investigation, "evolution" denotes the class of processes through which more complex systems arise from less complex ones. Evolution is thus the total of processes of organisational differentiation (von Bertalanffy 1968, p. 70; Nicolis and Prigogine 1977, Part V; Haken 1978, Chap. 10). This definition needs qualification for several reasons. In modern biology, with which the following chapters are largely concerned, processes of differentiation are often subsumed under the concept of development and strictly distinguished from evolutionary processes. Organic evolution, on its part, has almost universally been approached within the neo-Darwinian conceptual framework of natural selection and adaptation. As has been emphasised by Williams (1974, pp. 34–55) and others, natural selection and adaptation by no means correlate consistently with increasing organic complexity in natural history (but cf. Bonner 1988). On the other hand, organic evolution is clearly connected with a variety of phenomena such as ecological complexity, the origins of life, and human sociocultural differentiation, which are notoriously difficult to explain in evolutionary terms because they pose highly non-trivial problems of natural self-organisation, leading far beyond the domains of neo-Darwinian theory. It is these problems on which the present investigation concentrates. In any case, it will be evident from the context, when other connotations of "evolution," too, are intended. These connotations may then be indicated by the phrases "adaptive evolution", "neo-Darwinian evolution", and the like.

1.1 Examples

Before introducing explicit definitions, we illustrate briefly what we understand by an evolutionary process here. Our first example is chosen from population biology. Let $X_1, X_2, \ldots$ be a sequence of numbers for which

$$X_{n+1} = X_n \exp(k(1 - X_n/K)), \quad n = 1, 2, \ldots$$

holds. The latter equation is well known to describe a simple ecological mechanism of population growth ("logistic growth") in discrete time (non-overlapping generations in the population); its continuous-time analogue will be repeatedly considered in later sections. X_n is the population number after n generations ($X_n \geq 0$), K is the ecological carrying capacity of the environment, and k the rate constant of population growth. The above equation states that populations below the carrying capacity ($X_n \leq K$) grow logistically, that is, the growth rate decreases with X_n. Now it is well known to theoretical population biologists that the asymptotic behaviour of X_n

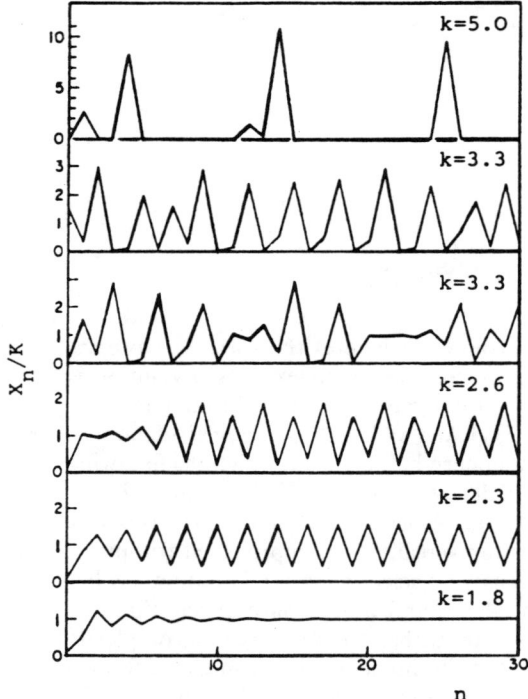

Fig. II.1. Different forms of discrete-time logistic population growth depending on the rate constant k, with normalised population number X/K as a function of n generations. The two cases with $k = 3.3$ are distinct by initial value. (May 1975)

depends sensitively on k (May 1975, 1976). Figure II.1 illustrates this dependence, showing increasingly regular behaviour of X_n as k decreases. Accordingly, a decrease in k causes the pattern of population growth to evolve from irregular to more ordered sequences of population numbers, whereby the rise of more ordered patterns from disordered ones constitutes an increase in organisational complexity.

Our second example is familiar from physics. Figure II.2a shows a ferromagnetic bar with magnetic field lines and complete parallel order of elementary (atomic) magnets. If the bar is heated up above a charateristic temperature T_c, called the ferromagnetic Curie temperature, the previous order of the elementary magnetic dipoles and, hence, the overall magnetisation of the bar vanish as shown in Fig. II.2c. When the bar is cooled down below T_c, it resumes its magnetisation steadily, corresponding to Fig. II.2b.

Figure II.3 shows the degree of magnetisation $\mathcal{M}$ of the bar as a function of the temperature. For temperatures far below T_c the proportion of elementary diples that are lined up is close to unity. This situation corresponds to an absolute maximum of $\mathcal{M}$. For temperatures above, and equal to, T_c the magnetisation is zero, corresponding to random angular distribution of the atomic magnets depicted in Fig. II.2c. In the intermediate temperature regime, the alignment of the elementary dipoles is incomplete, with intermediate values of $\mathcal{M}$.

Since $\mathcal{M}$ is a property of the bar as against the magnetic properties of the bar's elementary dipoles, Figs. II.2 and II.3 correspond respectively to microscopic and macroscopic descriptions of temperature-dependent transitions in the magnetic state. At the microscopic level, decreases in temperature within the subcritical regime are

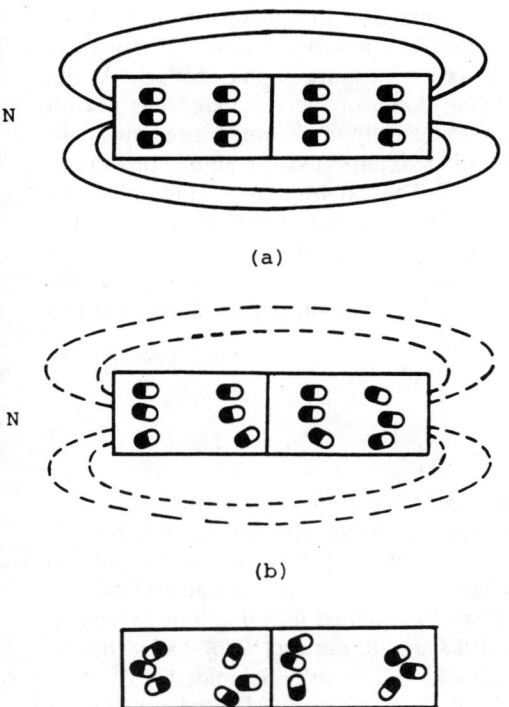

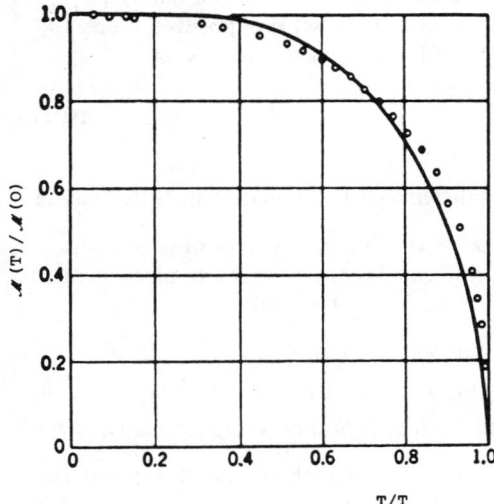

Fig. II.2a–c. Ferromagnetic bars at various degrees of magnetisation. **a** Complete alignment of atomic dipoles, with strong magnetic field outside. **b** Partially, and **c** completely, random angular distribution of atomic dipoles, with intermediate and zero outside field respectively. N and S indicate magnetic north and south poles respectively

Fig. II.3. Magnetisation $\mathscr{M}$ of nickel as a function of temperature T, with T_c being the so-called ferromagnetic Curie temperature (after Haken 1978)

evolutionary processes because they involve increases in coherent behaviour, or complexity, due to increasing dipole alignment. In other words, the irregular distribution of atomic dipoles is then replaced by a more regular pattern. At the macroscopic level of description, the cooling down of the bar magnet is not an evolutionary process. It is characterised exclusively by temperature-dependent variations in $\mathcal{M}$, and no spatio-temporally coherent patterns form among previously non-interacting objects *in this representation*. The example also shows that the evolutionary interpretation which a process can be given generally depends on the particular concepts, laws and theories in terms of which the process is described. Ignorance of this elementary fact feeds much useless controversy in current evolutionary debates. We begin the analysis of this situation by introducing various definitions.

1.2 Basic Definitions

The concepts of structure and system have been given adequate formal representations in set-theoretic terms in the mathematical and system-theoretical literature. Part of the conceptual framework we are going to develop now can also be found in Pichler (1975) and Mesarović and Takahara (1975), with occasional deviation in formal exposition and content. For less formally oriented introductions to general system theory we refer to von Bertalanffy (1968) and Rapoport (1986). As for the set-theoretic apparatus, most elementary textbooks will do (e.g., Schmidt 1973).

Let M be a non-empty set. Then R is an *n-place relation in M* if and only if n is a positive integer and $R \subset M^n$, where the symbol "M^n" denotes the n-fold Cartesian product of M, and n is called the *multiplicity* of R. Furthermore, R is called a *relation in M* just in case there exists a positive integer n so that $R \subset M^n$. We also use the following representation for relations. Let J be a non-empty set, but arbitrary otherwise. Let further $\mathcal{A}$ be a set and φ a function mapping J into $\mathcal{A}$, i.e., $\varphi: J \to \mathcal{A}$. The range of φ is then called a *family* and is written as $[A_j]_{j \in J}$, where $\varphi(j) = A_j$, $A_j \in \mathcal{A}$ for every j, $j \in J$. The domain J of φ is termed the *index set* of the family $[A_j]_{j \in J}$. Evidently, if J is finite (i.e., there is a finite subset J' of the set $\mathbb{P}$ of positive integers so that J and J' are isomorphic) and $[A_j]_{j \in J}$ is a family of sets, every R with

$$R \subset \underset{j \in J}{\times} A_j \tag{II.1}$$

is a relation in $\bigcup_{j \in J} A_j$. Similarly every set R satisfying (II.1) is a relation in M if J is finite and $[A_j]_{j \in J}$ is a subset of $\mathcal{P}(M)$, where "$\mathcal{P}$" is the familiar symbol of power sets.

The elements of an n-place relation R in M are n-tuples (n-place arrays, sequences) of elements of M in the usual sense. The set K_k with

$$K_k = \{x \mid \vee x_1 \cdots \vee x_{k-1} \vee x_{k+1} \cdots \vee x_n (\langle x_1, \ldots, x_{k-1}, x, x_{k+1}, \ldots, x_n \rangle \in R)\}$$

for $1 < k < n$ and analogous definitions for $k = 1$ and $k = n$, are called the *k-th component*, $[K_k]_{k \leq n}$ the *decomposition*, and $\bigcup_{k \leq n} K_k$ the *field* of R. Occasionally, when misunderstandings cannot arise, we also refer to the members of an n-tuple as the components of this n-tuple. Finally, one-place relations are sometimes called *properties*, *n*-place relations with $n \geq 2$ are also called *multiple relations*, and instead of the term "relation" we may use "attribute" in the following sense. If x is an element of the field of a relation R, we may say that x has the attribute R.

For every M and S, the ordered pair $\langle M, S \rangle$ is termed a *structure* if and only if M is a non-empty set and S is a non-empty set of relations in M. The set M is then called the *universe*, or *basic set*, of the structure $\langle M, S \rangle$. Note that except for $S \neq \emptyset$ the cardinality of S is not subject to further restriction in this definition. Moreover, we shall also meet with more complicated structures such as arise from relations in sets of relations. In order to minimise the difficulties in their explicit representation, we introduce the following abbreviations and definitions.

Let $\langle M, S \rangle$ be a structure, k a positive integer and C a relation with $C \subset S^k$. We call C *homogeneous in S* if and only if there are positive integers $n_1, \ldots, n_k$ so that

$$C \subset \mathscr{P}(M^{n_1}) \times \cdots \times \mathscr{P}(M^{n_k}), \tag{II.2}$$

and *inhomogeneous in S* otherwise. Evidently, if C is homogeneous in S, for every sequence $R_1, \ldots, R_k$ with

$$\langle R_1, \ldots, R_k \rangle \in C, \tag{II.3}$$

R_l is an n_l-place relation in M, $l \leq k$. Below we concern ourselves with an important class of inhomogeneous relations in a set S^* with $S \subset S^*$; this class consists of the so-called coupling relations defined for input-output systems with structure in $\langle M, S^* \rangle$. The relevant terminology will be introduced below. The inhomogeneity of such a coupling relation C is due to the fact that the systems for which it is defined are relations in M with arbitrary finite multiplicities so that (II.2) does not hold for it. Nonetheless we need not consider inhomogeneous relations in S explicitly here because every inhomogeneous relation in S can clearly be partitioned into a class of pairwise disjoint subrelations that are homogeneous in S.

It is possible to transform the relation C in (II.2) into a uniquely determined relation Q in $\mathscr{P}(M)$ with

$$Q \subset (\mathscr{P}(M))^{n_1 + \cdots + n_k}. \tag{II.4}$$

The transformation aimed at can be conveniently constructed in several steps. Let the symbol "$\rangle\langle$" denote the function which assigns the m-tuple $\rangle z\langle = \langle \{x_1\}, \ldots, \{x_m\} \rangle$ to the m-tuple $z = \langle x_1, \ldots, x_m \rangle$. For every m-place relation R in M define

$$\Phi(R) = \{y | \vee z(z \in R \wedge y = \rangle z\langle)\}. \tag{II.5}$$

Clearly, $\Phi(R) \subset (\mathscr{P}(M))^m$. Assume now that W is the set of relations in $\mathscr{P}(M)$ which have the form (II.5),

$$W = \{X | \vee m \vee R(m \in \mathbb{P} \wedge R \subset M^m \wedge X = \Phi(R))\}.$$

Then Φ is a one-to-one mapping from the set of relations in M onto W. Hence, if we introduce the definition

$$\wedge u(u \in U \Leftrightarrow \vee R_1 \cdots \vee R_k(\langle R_1, \ldots, R_k \rangle \in C \wedge u = \langle \Phi(R_1), \ldots, \Phi(R_k) \rangle)), \tag{II.6}$$

the relations C and U are isomorphic.

The final step in the reconstruction of (II.4) consists in the definition of a relation Q in $\mathscr{P}(M)$ in terms of U. Let the symbol "$[]$" denote the function assigning the

Cartesian product $\underset{l \leq m}{\times} A_l$ to the m-tuple of sets $\langle A_1, \ldots, A_m \rangle$. Now put

$$Q = \bigcup_{u \in U} [\![u]\!]. \tag{II.7}$$

Since by construction for every u, $u \in U$, one has

$$[\![u]\!] \subset (\mathcal{P}(M))^{n_1 + \cdots + n_k},$$

Q indeed satisfies (II.4). One also has

$$\wedge R_1 \cdots \wedge R_k (\langle R_1, \ldots, R_k \rangle \in C \Rightarrow \Phi(R_1) \times \cdots \times \Phi(R_k) \subset Q). \tag{II.8}$$

The relations C and Q are neither isomorphic (difference in cardinality between C and U, on the one hand, and Q, on the other), nor can the implication in (II.8) be reversed. If the latter were the case, all expressions of the type (II.3) could be replaced by inclusions of the type $\Phi(R_1) \times \cdots \times \Phi(R_k) \subset Q$ in all contexts, i.e., by expressions which contain at most relations in $\mathcal{P}(M)$. Unfortunately this is impossible, and yet, the definition (II.7) turns out to be very useful. Later we wish to consider theories semantically interpreted relative to structures whose universes are themselves sets of relations, and whose relations have the form (II.2). Assume T is such a theory, C a relation of the form (II.2) and C falls within the range of semantic interpretation of T. By use of the transformation Φ and definition (II.6), T can be given another interpretation in terms of relations similar in form to U, where C and U correspond as in (II.6). The old and the new interpretation are thus isomorphic. In Section III.3 we then show that the new interpretation can, on its part, be replaced as follows. The relations associated with it are transformed according to (II.7). The theory T then goes over into a theory T' with roughly the following properties. The descriptive languages of the two theories are connected in such a way that, if U is the extension of a concept of T, Q is the extension of a concept of T', where U and Q correspond as in (II.7). The semantic structure of T is transformed into a semantic interpretation of T' under which T' is true. Thus T' bears the entire material content of T.

Anticipating these results, we need not mind explicit representations of relations of the type (II.2), but may concern ourselves exclusively with the transformed relations (II.7) instead. This simplifies the formal apparatus of the present investigation to a great extent.

We are now prepared to define the important concept of hierarchical structure. An n-tuple $\Sigma = \langle \langle M_1, S_1 \rangle, \ldots, \langle M_n, S_n \rangle \rangle$ is called an *n-level hierarchy of structures*, or, more casually, though conveniently, an *n-level hierarchical structure* if and only if

1. $\langle M_1, S_1 \rangle$ is a one-level hierarchical structure, i.e., a structure; and in case $n \geq 2$
2. $\langle \langle M_1, S_1 \rangle, \ldots, \langle M_{n-1}, S_{n-1} \rangle \rangle$ is an $(n-1)$-level hierarchical structure;
3. $M_n \subset \mathcal{P}(M_{n-1})$; and
4. for every Q, $Q \in S_n$, there is an integer k, $k \geq 1$, and a relation C, $C \subset S_{n-1}^k$, so that C is homogeneous in S_{n-1} and

$$Q = \bigcup_{u \in U} [\![u]\!], \tag{II.7'}$$

where

$$U = \{u | \vee R_1 \cdots \vee R_k (\langle R_1, \ldots, R_k \rangle \in C \wedge u = \langle \Phi(R_1), \ldots, \Phi(R_k) \rangle)\}. \tag{II.6'}$$

It is follows immediately that for every r, $2 \leq r \leq n$, $M_r \subset \mathcal{P}(M_{r-1})$. Similarly, for every positive integer r, $r \leq n$, $\langle M_r, S_r \rangle$ is a structure. The structure $\langle M_r, S_r \rangle$ is then referred to as the *rth level of organisation in* Σ, or, equivalently, *stratum of rank r in* Σ.

The theoretical signficance of the concept of hierarchical structure rests upon the fact that one of the most prominent features of the structure of matter, as investigated in the physical and in the life sciences, is hierarchical organisation. We deal with this issue extensively below.

Historically, the notion of system, as used in 20th century empirical science, has a variety of roots ranging from theoretical biology to electronic engineering. In virtually all of these contexts, the notion has been intended to explicate intuitive views of wholes, or compound entities, arising from stable or recurrent relationships between elementary entities, or parts. Accordingly, systems as "complexes of elements standing in interaction" (von Bertalanffy 1968, p. 33) have been given formal representations simply as relations in mathematical system theory. In the present investigation, we are often concerned, at one and the same time, with various kinds of systems defined on quite different sets, for instance, systems of objects and systems of these systems of objects, and the like. The following definition proves reasonably appropriate for handling this situation.

We call s a *system with structure* $\langle M, S \rangle$ if and only if

1. $M \neq \emptyset$;
2. $S = [R_j]_{j \in J}$, where $[R_j]_{j \in J}$ is a finite family of relations in M (i.e., J is a finite index set); and
3. $S \subset \underset{j \in J}{\times} R_j$.

Furthermore, s is called a *system* if and only if there is a pair $\langle M, S \rangle$ so that the conditions (1) to (3) are satisfied. Evidently, if s is a system with structure $\langle M, S \rangle$, s is a relation in M. Let $\langle M, S \rangle$ and $\langle M, S^* \rangle$ be structures, $S \subset S^*$, and s a system with structure $\langle M, S \rangle$. Then we also say that s is a *system with structure in* $\langle M, S^* \rangle$.

The present concept of system and its definition relative to the concept of structure are more advantageous with respect to application than the usual definition of systems as relations proper. The concept of system introduced here clearly extends to relations between arrays of different length (i.e., number of components), which gives more flexibility in the explicit representation of multiple interactions between these objects.

We call a system s with structure $\langle M, S \rangle$ an *input–output system (binary system)* if and only if S has exactly two elements. They are respectively referred to as the *input* and *output relations*, or *sets*, of s. The components of the input and output relations of a binary system are the *input* and *output components*, respectively, of this system.

The concept of input–output system is used in science to describe multiple assignments of responses (outputs) to environmental cues (inputs) as are made by organisms, machines and information processing devices with complicated internal organisation. Input–output relations thus provide convenient representations of all sorts of physical, biological and other entities in terms of their observable behavioural attributes. In particular, the concept of open system, subsuming thermodynamic systems that exchange matter, energy and entropy with their environments, has been explicated with reference to input–output systems in biophysics, theoretical biology and information theory.

For later use we introduce the concept of state. Let s be a system with structure

$\langle M, S \rangle$ and P a non-empty set. A function $\varphi: P \to \mathscr{P}(s)$ is called a *state parametrisation of s* if and only if $\bigcup_{p \in P} \varphi(p) = s$. Sometimes the range $[\varphi(p)]_{p \in P}$ of φ rather than φ itself is referred to as the state parametrisation of s for the sake of brevity. Adopting the mathematical terminology of the theory of systems of differential equations, the domain P is also called a *state space*, or *parameter space, for s*, its elements being the *states of s*.

In general system theory, the concept of coupling, or interconnexion, of systems is intended as a global characterisation of the extreme diversity of interactions among physical, biological, etc. systems. In the most general sense, couplings of two or more systems may therefore be defined as multiple relations in suitably chosen sets of systems. The following definition is restricted to input–output systems, however, because in later chapters we concern ourselves primarily with binary systems. Given a sequence $s_1, \ldots, s_k$ of input–output systems with respective structures $\langle M, S_1 \rangle, \ldots, \langle M, S_k \rangle$, where $k \geq 2$. Thus, for every $l, l \leq k$, there are relations R_{l1} and R_{l2} in M with $S_l = \{R_{l1}, R_{l2}\}$, and positive integers m_l, n_l so that the decompositions of R_{l1} and R_{l2} are $[X_{li}]_{i \leq m_l}$ and $[Y_{lj}]_{j \leq n_l}$, respectively. The families $[X_{li}]_{l \leq k, i \leq m_l}$ and $[Y_{lj}]_{l \leq k, j \leq n_l}$ are then referred to as the *joint decomposition of* the input and output relations, respectively, of $s_1, \ldots, s_k$.

A system s is a (input–output) *coupling*, or *connexion, of* $s_1, \ldots, s_k$ if and only if

1. there are index sets I and J, card $I = m_1 + \cdots + m_k$, card $J = n_1 + \cdots + n_k$, and families of sets $[X_i]_{i \in I}$ and $[Y_j]_{j \in J}$ so that $[X_i]_{i \in I}$ is the joint decomposition of the input relations, and $[Y_j]_{j \in J}$ the joint decomposition of the output relations, of $s_1, \ldots, s_k$;
2. there are subsets $I^* \subset I$ and $J^* \subset J$ with $I^* = I$ if and only if $J^* = J$ so that,
3. when $R_1 = \underset{i \in I^*}{\times} X_i$ and $R_2 = \underset{j \in J^*}{\times} Y_j$, s is a system with structure $\langle M, S^* \rangle$, where $S^* = \{R_1, R_2\}$;
4. for every $j, j \in J \setminus J^*$, there is an index $i, i \in I \setminus I^*$, such that
$$X_i \cap Y_j \neq \emptyset,$$
and conversely,
5. for every $i, i \in I \setminus I^*$, there exists some $j, j \in J \setminus J^*$, so that
$$X_i \cap Y_j \neq \emptyset.$$

Although this definition looks rather unwieldy, the meaning of the concept defined is easy to grasp. Condition (3) states that the input relation R_1 and the output relation R_2 of the coupled system s are Cartesian products of components selected from the decompositions of the input and output relations, respectively, of $s_1, \ldots, s_k$. Conditions (3) to (5) demand that in case $I^* \neq I$ and $J^* \neq J$ there exist mixed pairs of intersecting input and output components not occurring in R_1 and R_2. Intersection of these remaining components means that the system to which they belong can be "wired" among one another, that is, some of the systems may feed others with their outputs. The following examples are schematically depicted in Fig. II.4.

(a) For every $l, l \leq k$ and $k \in \mathbb{P}$, let $s_l \subset R_{l1} \times R_{l2}$. Define
$$\underset{l \leq k}{\times} s_l = \{\langle x_1, \ldots, x_k, y_1, \ldots, y_k \rangle | \langle x_1, y_1 \rangle \in s_1 \wedge \cdots \wedge \langle x_k, y_k \rangle \in s_k\}.$$

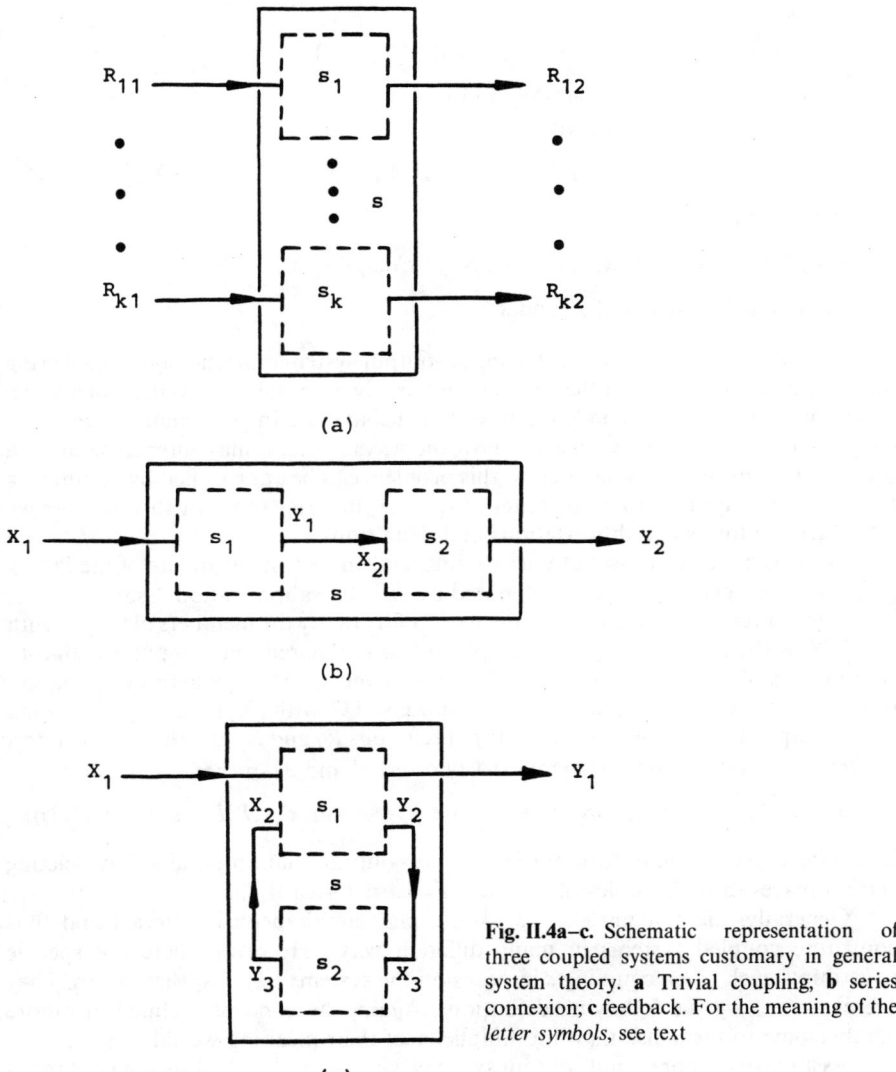

Fig. II.4a–c. Schematic representation of three coupled systems customary in general system theory. **a** Trivial coupling; **b** series connexion; **c** feedback. For the meaning of the letter symbols, see text

The system

$$s = \underset{l \leq k}{\bigtimes} s_l \subset \left(\underset{l \leq k}{\times} R_{l1} \right) \times \left(\underset{l \leq k}{\times} R_{l2} \right)$$

is called the *trivial coupling* of $s_1, \ldots, s_k$.

(b) Given two systems

$$s_1 \subset X_1 \times Y_1, \quad s_2 \subset X_2 \times Y_2, \quad X_2 \cap Y_1 \neq \emptyset.$$

The system
$$s = \{\langle x,y \rangle | \vee z(\langle x,z \rangle \in s_1 \wedge \langle z,y \rangle \in s_2)\} \subset X_1 \times Y_2$$
is called a *series connexion* of s_1 and s_2.

(c) Given two systems s_1, s_2 with
$$s_1 \subset X_1 \times X_2 \times Y_1 \times Y_2, \quad s_2 \subset X_3 \times Y_3, \quad X_2 \cap Y_3 \neq \emptyset, \quad X_3 \cap Y_2 \neq \emptyset.$$
The system
$$s = \{\langle w,y \rangle | \vee x \vee z(\langle w,x,y,z \rangle \in s_1 \wedge \langle z,x \rangle \in s_2)\} \subset X_1 \times Y_1$$
is a *feedback coupling* of s_1 and s_2.

In applications of the concept of input–output system in science and engineering one is often concerned with the problem of how the internal organisation of a given system influences this system's input–output behaviour. In particular, when s is a coupling of $s_1, \ldots, s_k$, one wishes to know the ways $s_1, \ldots, s_k$ may interact with each other so as to produce s. Analytically, this problem can be approached by attempting to express the input–output attributes of $s_1, \ldots, s_k$ in terms of s. For this purpose we introduce the following abbreviations and definitions.

Let I be an arbitrary set of positive integers, and let the members of the family $[x_i]_{i \in I}$ be arranged in a sequence denoted by "x". The abbreviation "ssq(x', x, I', I)" is used to express that x' is the subsequence of x formed by the members of $[x_i]_{i \in I'}$ with $I' \subset I$. Now the concept of system component, as compared to the input and output components of a system, is defined as follows. Given s, I and J, where s is an input–output system and both I and J are finite subsets of $\mathbb{P}$, with $[X_i]_{i \in I}$ and $[Y_j]_{j \in J}$ being the decompositions of the input and output relations R_1 and R_2 of s, then s' is a *system component of s* if and only if there exist two sets I' and J' so that
$$s' = \{\langle x',y' \rangle | \vee x \vee y(x \in R_1 \wedge y \in R_2 \wedge \langle x,y \rangle \in s \wedge \text{ssq}(x',x,I',I) \wedge \text{ssq}(y',y,J',J))\}.$$

Roughly, a system component of s is an input–output relation obtained by deleting certain places in the n-tuples of s, where $n = \text{card } I + \text{card } J$.

Generally, the systems $s_1, \ldots, s_k$, $k \geq 2$, may simultaneously interact and thus constitute coupled systems in many different ways. However, there are specific constraints each given coupling s imposes on the systems $s_1, \ldots, s_k$ themselves. They are expressed by the following definitions. Again, the concepts defined are more cumbersome to formalise than the simplicity of their meaning would suggest.

Assume once more input–output systems, $s, s_1, \ldots, s_k$, $k \geq 2$, of such a kind that s is a non-trivial coupling of $s_1, \ldots, s_k$. Let the families $[X_i]_{i \in I}$ and $[Y_j]_{j \in J}$ be the joint decompositions of the input and output relations R_{l1} and R_{l2}, respectively, of $s_1, \ldots, s_k$, where $l \leq k$. Furthermore, assume index sets I^* and J^*, $I^* \subset I$, $J^* \subset J$, so that the family $[X_i]_{i \in I^*}$ is the decomposition of the input relation R_1, and $[Y_j]_{j \in J^*}$ the decomposition of the output relation R_2, of s. Put

$$s^* = \{\langle x^*, y^* \rangle | \vee x \vee y(x \in \underset{l \leq k}{\times} R_{l1} \wedge y \in \underset{l \leq k}{\times} R_{l2} \wedge \langle x,y \rangle \in \underset{l \leq k}{\times} s_l \wedge \text{ssq}(x^*, x, I^*, I)$$
$$\wedge \text{ssq}(y^*, y, J^*, J))\}$$

Then the systems $s_1, \ldots, s_k$ are referred to as *coupled subsystems of s* if and only if

$$s^* = s, \qquad \qquad \text{(II.9)}$$

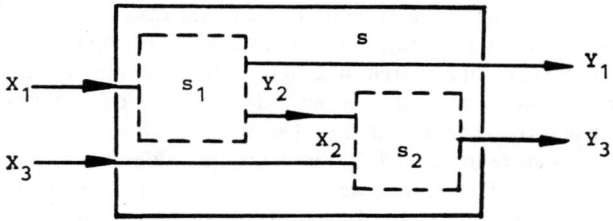

Fig. II.5. Cascade connexion s of the systems s_1 and s_2

and *decoupled subsystems of* s otherwise. (Mind the difference between *subsystems* of a system s and *coupled*, or *decoupled*, *subsystems* of s. By the former we simply mean subrelations of s. We do not expect difficulties to arise from this slightly ambiguous terminology which accords at least with major parts of the system-theoretical literature.)

The meaning of the above definitions can be circumscribed briefly as follows. By construction, s^* is a system component of the trivial connexion $\underset{l \leq k}{\times}\, s_l$. The input and output components of s^* have the same indices as the input and output components of s. If (II.9) holds, there are system components of $s_1, \ldots, s_k$ uniquely determining the environmental interactions of the coupled system. The remaining components then constitute the internal organisation of s in the sense that every connexion of these components among each other is compatible with the input–output behaviour of the entire system s. We shall use the distinction between coupled and decoupled subsystems to explicate intuitive notions of part-whole relationships. When the coupled system s is viewed as a whole, and $s_1, \ldots, s_k$ are its parts, the parts are conceived of as being given before and after their assemblage, depending on whether $s_1, \ldots, s_k$ are decoupled or coupled subsystems of s.

The following example is visualised in Fig. II.5. Let

$$s_1 \subset X_1 \times Y_1 \times Y_2, \quad s_2 \subset X_2 \times X_3 \times Y_3, \quad X_2 \cap Y_2 \neq \emptyset,$$

with $[X_i]_{i \leq 3}$, $[Y_j]_{j \leq 3}$ being the joint decompositions of the input and output relations of s_1 and s_2. Define the coupling

$$s = \{\langle w, x, y, z \rangle \mid \vee u (\langle w, y, u \rangle \in s_1 \wedge \langle u, x, z \rangle \in s_2)\}$$

of s_1 and s_2 (known as the *cascade connexion*). Then s_1 and s_2 are coupled subsystems of s if, for instance, $s_1 = X_1 \times Y_1 \times (X_2 \cap Y_2)$ and $s_2 = (X_2 \cap Y_2) \times X_3 \times Y_3$. Conversely, s_1 and s_2 are decoupled subsystems of s if, as an example, $s_1 = X_1 \times Y_1 \times Y_2$, $s_2 = X_2 \times X_3 \times Y_3$ and $X_2 \neq Y_2$.

Like the concept of hierarchical structure, notions of hierarchical system play a major role in the system – theoretical literature (von Bertalanffy 1968; Mesarović et al. 1970; see also Causey 1977, Chap. 7, and the references therein). Here we call a system with the structure $\langle M, S \rangle$ an *n-level hierarchical system* if and only if there exists an n-level hierarchy of structures $\langle \langle M_1, S_1 \rangle, \ldots, \langle M_n, S_n \rangle \rangle$ so that $\langle M, S \rangle = \langle M_n, S_n \rangle$. Instead of "$n$-level hierarchical system" we also use the expression "n-level stratified system", omitting the adjunct "n-level", however, whenever it is irrelevant to the context.

Hierarchical stratification is characteristic primarily of systems that can be described at distinct (e.g., microphysical and macrophysical) levels of abstraction.

Examples frequently referred to in the context of the hierarchical organisation of matter are crystals formed from atoms and molecules, organisms consisting of cells, and social groups forming on the basis of recurrent interactions between individuals. We briefly illustrate the connexion between microscopic and macroscopic representations of stratified systems, using two-level hierarchies. Let $\langle M_1, S_1 \rangle$ be a structure and s_1 a binary system with structure in $\langle M_1, S_1 \rangle$. There are thus relations R_1, R_2 with $R_1 \in S_1, R_2 \in S_1$ so that $s_1 \subset R_1 \times R_2$. From s_1 we construct a system s_2, replacing ordered pairs of elements of R_1 and R_2 by ordered pairs of subsets of R_1 and R_2 (cf. Mesarović and Takahara 1975, p. 257). This procedure suggests itself in applications in which the inputs and outputs of systems cannot be specified (observed, etc.) uniquely relative to the respective universes. However, in such cases it may still be possible to delineate subsets of R_1 and R_2, and specify relations between them, which constitute the relevant observable attributes of the system concerned. Accordingly, if

$$s_2 \subset \{\langle X, Y \rangle | X \subset \{x | x \in R_1 \wedge \vee z(\langle x, z \rangle \in s_1)\} \wedge Y = \{y | \vee x(x \in X \wedge \langle x, y \rangle \in s_1)\}\},$$

we call s_2 a *macrosystem* associated with s_1, and s_1 a *microsystem* associated with s_2. By construction, $s_2 \subset \mathscr{P}(R_1) \times \mathscr{P}(R_2)$. Using the framework of (II.2) to (II.7), one straightforwardly transforms s_2 into a relation s'_2 in $\mathscr{P}(M_1)$. Defining

$$Q_i = \Phi(R_i), \quad i = 1, 2,$$

and assuming the function Φ to be given by (II.5), one verifies that $s'_2 \subset Q_1 \times Q_2$ and that $\langle\langle M_1, \{R_1, R_2\}\rangle, \langle \mathscr{P}(M_1), \{Q_1, Q_2\}\rangle\rangle$ is a two-level hierarchical structure.

In other applications, it may happen that the multiplicities of R_1 and R_2, which may read m_1 and m_2, respectively, are very large. From the macrophysical point of view, the structure of s_1 may then seem blurred, meaning that it may be difficult, or practically impossible, to determine the order of the input and output components of s_1. Denoting the fields of R_1 and R_2 by "F_1" and "F_2", one can then consider the system

$$s_2 = \{\langle X, Y \rangle | \vee x_1 \cdots \vee x_{m_1} \vee y_1 \cdots \vee y_{m_2}(\langle x_1, \ldots, x_{m_1}, y_1, \ldots, y_{m_2}\rangle \in s_1$$
$$\wedge X = \{x_1, \ldots, x_{m_1}\} \wedge Y = \{y_1, \ldots, y_{m_2}\})\}$$

instead of s_1, with $s_2 \subset \mathscr{P}(F_1) \times \mathscr{P}(F_2)$. Hence, s_2 is a two-level stratified system, the hierarchy of structures $\langle\langle M_1, \{F_1, F_2\}\rangle, \langle \mathscr{P}(M_1), \{\mathscr{P}(F_1), \mathscr{P}(F_2)\}\rangle\rangle$ being trivially associated with it. Once more, it seems intuitively plausible to call s_2 a *macrosystem* associated with s_1, and s_1 a *microsystem* associated with s_2.

Another important connotation of the concept of hierarchical system is rank order in levels of control, or decision-making, in systems that are called functionally differentiated by biologists and social scientists. As an example, consider governmental control of subordinate bureaucracies. The behaviour of the subordinate administrations is characterised by the conversion of inputs, such as information, taxes, applications by the clientele, etc., into outputs (decisions, regulations, expenses, etc.). Governmental decision-making as a response to the performance of the subordinate administrative units can now be viewed as an input–output pattern acting so as to transform the modes of lower-level bureaucratic decision-making. This kind of sociopolitical control hierarchy clearly satisfies the present abstract definition of hierarchical system.

There is a correspondence between the coupling of systems and systems stratification that is of great importance to evolutionary theories of hierarchical organisation. Let $\langle M,S \rangle$ be a structure and V a set of systems with structures in $\langle M,S \rangle$. Observe that $\langle M, S \cup V \rangle$ is a structure as well. Let k be a positive integer and C, R_1 and R_2 homogeneous relations in V so that

1. $C \subset R_1 \times R_2$ and
2. for every sequence $\langle s, s_1, \ldots, s_k \rangle$ with $\langle s, s_1, \ldots, s_k \rangle \in C$, s is a coupling of $s_1, \ldots, s_k$, where $s \in R_1$ and $\langle s_1, \ldots, s_k \rangle \in R_2$.

Thus R_1 is a one-place relation, and C a coupling relation, in V. Once more using the framework of (II.2) to (II.7), C, R_1 and R_2 can be transformed respectively into relations Q_C, Q_1 and Q_2 in $\mathscr{P}(M)$. The transformed Q_C of the coupling relation C is a binary system with structure $\langle \mathscr{P}(M), \{Q_1, Q_2\} \rangle$ and two-level hierarchy of structures $\langle \langle M, S \cup V \rangle, \langle \mathscr{P}(M), \{Q_1, Q_2\} \rangle \rangle$. We refer to the second stratum in this hierarchy, and any other stratum $\langle \mathscr{P}(M), S' \rangle$ with $\{Q_1, Q_2\} \subset S'$, as a *stratum appropriate to* the coupling relation C.

This correspondence between the interconnexion and stratification of systems extends to multilevel hierarchies. As a simple example, let C, C_1 and C_2 be non-trivial, homogeneous coupling relations in V, and let $s_1, s_2, s_{11}, s_{12}, s_{21}, s_{22}$ be systems contained in V so that

$$\langle s_i, s_{i1}, s_{i2} \rangle \in C_i, \quad i = 1, 2.$$

In other words, s_i results from interconnecting s_{i1} and s_{i2}. Assume that the system $s, s \in V$, is a coupling of s_1 and s_2 with

$$\langle s, s_1, s_2 \rangle \in C.$$

Since s_1 and s_2 themselves are coupled systems, the mode of connecting s_{11}, s_{12}, s_{21} and s_{22} changes from C_1 and C_2 to C when s is built from s_1 and s_2 (see example in Fig. II.6). This transformation corresponds to another relation C' with $\langle C, C_1, C_2 \rangle \in C'$; C' is a coupling relation itself since C, C_1 and C_2 possess representations as input–output systems.

The preceding example motivates the following definitions. Let V be a set of input-output systems with structures in $\langle M, S \rangle$, and let $s \in V$. Then s is a *coupled system of coupling degree* d_c in V if and only if

1. for $d_c = 0$, s is not a non-trivial coupling of any systems contained in V,
2. for $d_c = 1$, there is an integer $k, k \geq 2$, and a sequence of systems $s_i \in V$, $1 \leq i \leq k$, so that
   a) s is a non-trivial coupling of $s_1, \ldots, s_k$,
   b) for every i, $1 \leq i \leq k$, s_i is a coupled system of zero coupling degree in V, and
3. for $d_c \geq 2$, there is an integer $l, l \geq 2$, and a sequence of systems $s'_j \in V$, $1 \leq j \leq l$, so that
   a) s is a non-trivial coupling of $s'_1, \ldots, s'_l$, and
   b) for every j, $1 \leq j \leq l$, s'_j is a coupled system of coupling degree $d_c - 1$ in V.

This defintion does not exclude the possibility that s has different coupling degrees in V because s may be composed of systems in V in many different ways. However, where necessary or expedient, V can be chosen sufficiently restricted so that the coupling degree of s in V is uniquely determined.

A hierarchical system C is called an *r-level hierarchy of coupling relations based on*

V if and only if there is a structure $\langle M, S \rangle$, an r-level hierarchy of structures with basic stratum $\langle M, S \rangle$, and an integer l, $l \geq 2$, so that

1. V is a set if input–output systems with structures in $\langle M, S \rangle$;
2. in the trivial case $r = 1$, C is an l-place relation in M with $C \in V$;
3. for $r = 2$, C is a $(l + 1)$-place, non-trivial, homogeneous coupling relation in V, with the second stratum in the hierarchy being appropriate to C, and
4. if $r \geq 3$, for every sequence $\langle C', C_1, \ldots, C_l \rangle \in C$ and every j, $1 \leq j \leq l$, C_j is an $(r - 1)$-level hierarchy of coupling relations based on V, the $(r - 1)$th stratum being appropriate to C_j.

One verifies easily by induction that if s is a coupled system of coupling degree d_c in V, there exists an input–output system C which is a $(d_c + 1)$-level hierarchy of coupling relations based on V. Figure II.6 illustrates this correspondence between hierarchical rank and coupling degree for a stratified system with $d_c = 2$.

Notions of structural complexity are much less straightforward to characterise than the concept of structure itself. The relevant definitions vary from *number of interacting units* (individuals, systems, species, and the like) to *entropy measures* of (dis-)order in thermodynamics and information theory, corresponding to the perspectives and needs of each particular discipline. The situation is especially unfavourable to approaches to phenomena of hierarchical differentiation and related processes of natural self-organisation. On the one hand, the (quantitative and qualitative) concepts of complexity used in the natural and social sciences tend to be level-specific in the sense that the respective theories refer to semantically distinct strata of physical, organic and sociocultural interactions (Mesarović et al. 1970, pp. 30ff). On the other hand, it is precisely a cross-level comparative concept of structural complexity that is required in many evolutionary analyses. Detailed explications of hierarchies of structures in the physical and organic world may indeed prove exceedingly cumbersome. In view of this situation, only a few basic considerations concerning complexity are carried out here. They may, however, suffice to present the main argument of this chapter.

We call two structures $\langle M_1, S_1 \rangle$ and $\langle M_2, S_2 \rangle$ *partially comparable* if and only if they have non-trivial substructures $\langle M, S_1' \rangle$, $\langle M, S_2' \rangle$ with common universe $M = M_1 \cap M_2$, i.e.,

1. $M \neq \emptyset$, (II.10a)

and for $i = 1, 2$,

2. $S_i' = \{R' \mid \vee n \vee R(R \subset M_i^n \wedge R \in S_i \wedge R' = M^n \cap R)\}$, (II.10b)

3. $S_i' \neq \{\emptyset\}$. (II.10c)

Otherwise, the two structures are called *incomparable*, or *disparate*. They are termed *completely comparable*, or, briefly, *comparable* if and only if they are partially comparable and

4. $S_i' = S_i$, $i = 1, 2$. (II.10d)

Finally, they are *essentially comparable* if and only if they are partially comparable and $\langle M_1, S_2 \rangle$ or $\langle M_2, S_1 \rangle$ is a structure, too. In this case, namely, one of the

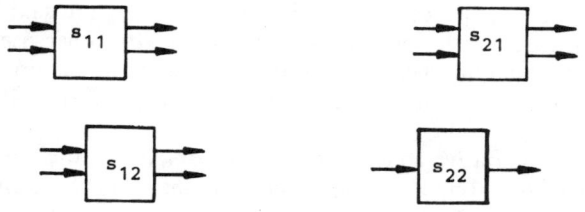

(a) Decoupled subsystems

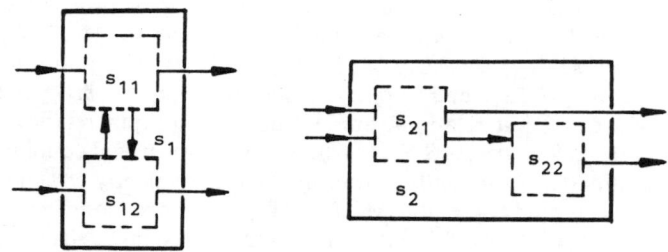

(b) s_1 as a double feedback coupling of s_{11} and s_{12}, and s_2 as a series connexion of s_{21} and s_{22}; in either case $d_c = 1$ in V.

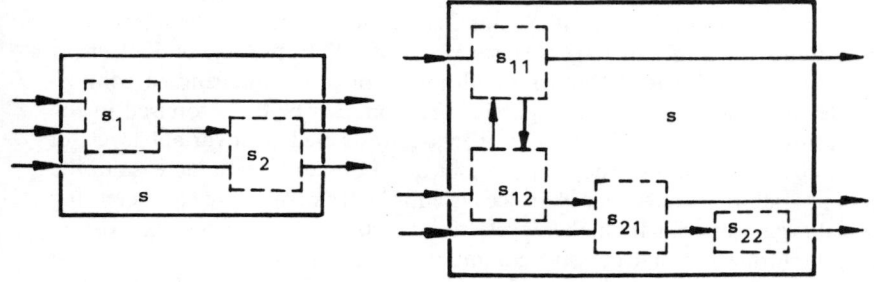

(c_1) s as a cascade connexion of s_1 and s_2.
(c_2) s as a multiple connexion of s_{11}, s_{12}, s_{21} and s_{22} of coupling degree $d_c = 2$ in V.

Fig. II.6a–c. Correspondence between coupling degree and hierarchical rank in stratified systems: example with $d_c = 2$. The step from **b** to **c** is due to a transformation operating on coupling relations in $V = \{s_{11}, s_{12}, s_{21}, s_{22}, s_1, s_2, s\}$. This transformation constitutes a three-level hierarchy of coupling relations

structures admits a trivial extension which is completely comparable with the other one.

The structure $\langle M_2, S_2 \rangle$ is said to be *more complex (evolved, differentiated) than* the structure $\langle M_1, S_1 \rangle$ if and only if

$$M_1 \subset M_2 \quad \text{and} \quad S_1 \subset S_2. \tag{II.11}$$

If (II.11), or, conversely, $M_2 \subset M_1$ and $S_2 \subset S_1$, the two structures are said to be *commensurable in complexity*, and *incommensurable in complexity* otherwise. If a structure is more complex than another one, the two are essentially comparable. Conversely, if two structures are incomparable, they are incommensurable in complexity.

We distinguish two comparative concepts of system complexity, one referring to the environmental interactions (input–output behaviour), the other to the internal organisation (coupled subsystems), of systems. A systems s_2 is said to be *behaviourally more complex (evolved) than* another system s_1 if and only if there exists a system component s of s_2 so that $s_1 \subset s$. This definition covers the possibility of $s = s_2 = s_1$, which we do not consider separately because in this case comparisons in system complexity become trivial anyway. Now assume two structures $\langle M_1, S_1 \rangle$ and $\langle M_2, S_2 \rangle$, and sequences $\langle s_1, s_{11}, \ldots, s_{1m_1} \rangle$, $\langle s_2, s_{21}, \ldots, s_{2m_2} \rangle$ of input–output systems with structures in $\langle M_1, S_1 \rangle$ and $\langle M_2, S_2 \rangle$ respectively ($m_1 \geq 2, m_2 \geq 2$). For $i = 1, 2$ let U_i be the set of input relations, and V_i the set of output relations, of $s_{i1}, \ldots, s_{im_i}$. Evidently, $U_i \subset S_i$ and $V_i \subset S_i$. Let further s_i be a non-trivial coupling of $s_{i1}, \ldots, s_{im_i}$. Then s_2 is said to be *structurally more evolved*, or, equivalently, *structurally more complex*, than s_1 if and only if $\langle M_2, U_2 \cup V_2 \rangle$ is more complex than $\langle M_1, U_1 \cup V_1 \rangle$. It follows immediately that every (coupled) system is structurally more complex than any of its (coupled or decoupled) subsystems. According to these definitions, greater structural complexity of coupled systems need not entail greater complexity with respect to input–output behaviour, and conversely. This independence of the two concepts of system complexity seems no more than intuitively plausible, however.

Clearly structure and complexity of the physical world can be theoretically analysed only in close connexion with semantic conventions such as the selection of interpretations of scientific languages (Mesarović et al. 1970, pp. 30ff; von Kutschera 1972, Sects. 4.7, 6.4). This gives rise to a multiplicity of context-dependent connotations of "evolution" which, in a rough classification, can be dichotomised in the following sense. There are evolutionary processes restricted to invariant levels of conceptual abstraction (e.g., structural diversification due to a mere increase in the number of interacting populations in an ecosystem). Alternatively, one has increasing hierarchical organisation, with higher levels of interaction forming from pre-existing ones (e.g., the formation of successive community levels in ecology). The following sections are mainly concerned with phenomena of the latter type for which a condition of advance in structural differentiation must be given within the present formalism. This amounts to the task of making the strata in a hierarchical structure comparable to one another.

Given the n-level hierarchical structure $\Sigma = \langle \langle M_1, S_1 \rangle, \ldots, \langle M_n, S_n \rangle \rangle$ for $n \geq 2$. Let the symbol "f" denote the function which, for every $r, 2 \leq r \leq n$, transforms S_r into

$$f(S_r) = \{W \mid \vee Q(Q \in S_r \wedge W = \bigcup_{q \in Q} [\![q]\!])\}. \tag{II.12}$$

Since for every m-place relation Q, $Q \in S_r$, and every $q, q \in Q$, one has $q \in (\mathscr{P}(M_{r-1}))^m$ and $[\![q]\!] \subset M_{r-1}^m$, $W = \bigcup_{q \in Q} [\![q]\!]$ is an m-place relation in M_{r-1} and $\langle M_{r-1}, f(S_r) \rangle$ a well-defined structure. With S_{r-1}^* being defined by

$$S_{r-1}^* = S_{r-1} \cup f(S_r), \tag{II.13}$$

the two structures $\langle M_{r-1}, S_{r-1} \rangle$ and $\langle M_{r-1}, S^*_{r-1} \rangle$ are trivially comparable, and the latter is more complex than the former.

Let $\Sigma = \langle\langle M_1, S_1 \rangle, \ldots, \langle M_m, S_m \rangle\rangle$ and $\Lambda = \langle\langle N_1, T_1 \rangle, \ldots, \langle N_n, T_n \rangle\rangle$ be m-level and n-level hierarchical structures, respectively. Generalising (II.13),

$$S^* = \bigcup_{r \leq m} f^{(r-1)}(S_r),$$

$$T^* = \bigcup_{r \leq n} f^{(r-1)}(T_r),$$

where "$f^{(r-1)}$" denotes $(r-1)$-fold iteration of f, one defines: Σ is *more complex*, or *evolved*, *than* Λ if and only if

$$N_1 \subset M_1 \quad \text{and} \quad T^* \subset S^*. \tag{II.14}$$

In order to exemplify the present definitions of structure and complexity in terms of concepts familiar elsewhere in science, a few remarks should be added. Assume M_1, M_2, S_1 and S_2 are sets such that $\langle M_1, S_1 \rangle, \langle M_2, S_2 \rangle$ are topological structures in the sense of mathematical topology theory (see, e.g., Grotemeyer 1969). One finds that if $\langle M_1, S_1 \rangle$ and $\langle M_2, S_2 \rangle$ are comparable, $\langle M, S_1 \rangle$ and $\langle M, S_2 \rangle$ are topological structures as well, were $M = M_1 \cap M_2 \neq \emptyset$. The inclusion $S_1 \subset S_2$ is then the condition of greater topological complexity ("refinement") of $\langle M, S_2 \rangle$.

Similarly, we briefly indicate the way the present comparative concept of complexity can be connected with the quantitative entropy concept which is widely used in science to assign degrees of complexity to structures and systems. The terminology, denotation and definitions we use are taken from Jones (1979). Let M be a sample space, with the assignment P of probabilities defined on it, and let S_1, S_2 be sets of events in M, i.e., $S_i \subset \mathcal{P}(M)$ for $i = 1, 2$. Suppose that

$$S_1 \subset S_2. \tag{II.11'}$$

Resuming the terminology of probability theory, the structure $\langle M, S_i \rangle$ is said to be realised as the outcome of an experiment (in a process, etc.) if and only if the event $E_i = \bigcap_{E \in S_i} E$ occurs in the experiment (process, etc.). Because of (II.11')

$$E_2 \subset E_1. \tag{II.11''}$$

Then

$$H(S_i) = -P(E_i) \log P(E_i) - (1 - P(E_i)) \log (1 - P(E_i))$$

is the entropy of the structure $\langle M, S_i \rangle$, where log is the logarithm (arbitrary base) and $i = 1, 2$. By use of conditional probabilities, the conditional entropy $H(S_2|S_1)$ can be defined as the entropy of $\langle M, S_2 \rangle$ given $\langle M, S_1 \rangle$ (i.e., provided E_1 has occurred) (Jones 1979, Chap. 2). It is then straightforward to show that

$$H(S_2|S_1) \leq H(S_1)$$

follows from (II.11''), with equality holding if and only if $E_1 = E_2$. In this simple case at least, structures that are more complex in the sense of (II.11) will have lower entropy, corresponding to the familiar notion of entropy as a measure of disorder.

Mesarović et al. (1970) also consider the important case of cross-level relations in hierarchies of structures. Such relations contribute to the complexity of the hierarchical structures concerned in a straightforward fashion. However, as we are primarily interested in logical aspects of evolutionary reasoning rather than the formalisation of multilevel orderings, we do not enter into these details here.

We briefly discuss a few more consequences of (II.14). First, it covers the case $\langle M_r, S_r \rangle = \langle N_r, T_r \rangle$, $r \leq n < m$, in which $T^* \subset S^*$ trivially holds. This means that adding further strata to a given hierarchy implies a trivial increase in complexity. If neither (II.14) nor the converse inclusions hold, Σ and Λ are incommensurable in complexity. In case $M_1 \cap N_1 = \emptyset$, $\langle M_1, S^* \rangle$ and $\langle N_1, T^* \rangle$ are not even comparable. These implications of (II.14), although apparently very restrictive, are reasonable, however, and may be illustrated by an example from evolutionary biology. Williams (1974, pp. 34ff) discusses a variety of instances of phylogenetic differentiation in molecular genetic, histological, morphological and taxonomic structure. For example, he relates to the popular pair of assumptions that, firstly, evolutionary advance from lower to higher taxa consists in increasing organic complexity and, secondly, that the change from fish to mammal exemplifies such progress. Although in some respects, such as brain structure, a mammal is certainly more complex than any fish, in other respects, such as integumentary histology, fish tend to be much more complex than any mammalian organism. The intuitively suggestive notion that mammals, when viewed as organic structures, are more evolved than fish is thus meaningless even if both structures have a common universe of cellular or biochemical objects. This notion corresponds exactly to the case that the commensurability condition (II.14) is not satisfied. On the other hand, if the analysis can be restricted to partially comparable neurophysiological substructures of the mammalian and fish brain, (II.14) may be fulfilled, and the statement "the mammalian brain is more evolved than that of fish" is indeed meaningful.

Relations (II.10) to (II.14) impose constraints on evolutionary theories of multilevel organisation which constitute the main topic of the following sections. It will be argued that in comparative analyses of *evolved* hierarchical complexity distinct levels of organisation must be made comparable and, furthermore, commensurable in order to assess evolutionary advance. More precisely, a common semantic frame of reference, that is, a common universe, and orderings like (II.11) and (II.14) must be assumed for any number of systems and strata distinguished by the observer. Otherwise hierarchies of structures cannot be understood as *evolutionary* successions in a well-defined sense. Many of the invalid or non-sensical views in current evolutionary thought with which the present chapter deals arise from violations of this postulate.

1.3 Informal Summary

It has been argued that many current evolutionary debates in science and philosophy remain inconclusive or even useless as long as the system-theoretic concepts referred to in these debates are not made sufficiently precise. By a *system* we understand any relation between the environmental stimuli (inputs) to which an object may be exposed, and the response (outputs) of this object. Input–output relations provide useful characterisations of all sorts of physical and biological objects in terms of their behavioural attributes, i.e., modes of environmental interaction. Accordingly, the present concept of *system structure* is strictly limited to sets of such behavioural attributes, whereas the *internal organisation* of a system is defined in terms of its *subsystems* and *coupling relations* between these subsystems as visualised in Figs. II.4

to II.6. Within the context of the present inquiry, the *hierarchical* organisation of structures and systems is of primary interest, whereby the term "hierarchical organisation" essentially refers to input–output relations (*macrosystems*) defined on sets of input–output relations (*microsystems*). Familiar examples of the hierarchical organisation of matter are crystals formed from atoms and molecules, organisms consisting of cells, and social groups arising from stable or recurrent interactions between individuals, with atoms, cells and individual organisms described in terms of their characteristic interactions.

A structure has been said to be *more complex* (*more evolved*) than another one if it includes this other one, i.e., contains additional attributes and relations. For any two systems, one is more complex (more evolved) than the other one if, roughly speaking, the structure of the former is more complex than that of the latter. Thus, in order to establish evolutionary relationships between any two systems, the structures of the systems must be (or must be made) *commensurable in complexity*. The situation is particularly intricate in comparative analyses of hierarchical complexity because in hierarchical systems the macro- and microsystems are relations defined on disjoint sets of objects and, hence, are themselves disjoint. It has been argued that structures and systems defined on disjoint sets of objects cannot be compared in complexity in a natural and straightforward way. In order to remedy the situation, a function has been construed which transforms the macrostructures of hierarchical systems into microstructures. In other words, it has been shown that hierarchical structures and systems can always be given uniquely defined microrepresentations to which the present comparative concepts of system complexity and evolution apply.

2 Basic Problems of the Evolution of Matter

Processes of structural diversification abound in the prebiotic and living world. According to modern cosmology, the universe has evolved from a homogeneous initial state of hot radiation ("big bang"). Adiabatic expansion of the universe caused the separation of matter from radiation, followed by the formation of galaxies, stars and, eventually, chemical elements. In the realm of organic life, genetic mutations and a variety of evolutionary mechanisms have established extreme ranges of biomolecular, organismic and ecological complexity.

However, structural differentiation is a type of process markedly distinct from other patterns of change in physical and organic systems. Earlier this century it was frequently maintained that evolution in the present sense even required new types of explanation beyond the theories of physics (e.g., teleological ones) because evolutionary processes were believed to involve crucial violations of the Second Law of Thermodynamics. This law states that isolated systems develop irreversibly towards thermodynamic equilibrium, that is, a state of maximum disorder and, equivalently, minimum structural coherence. Today it is obvious that this view of evolutionary processes arose from a misunderstanding of the premises of equilibrium thermodynamics. After the situation had become remedied by suitable theoretical approaches to non-equilibrium thermal processes, it turned out that the resulting scientific disciplines such as non-linear thermodynamics, (bio-)cybernetics and related fields have revealed novel and unexpected dimensions of self-regulation in prebiotic as well as organic systems.

Even with these improved theoretical tools at hand, the modern life sciences are still far from an adequate understanding of two fundamental evolutionary problems. These are the origins of life and the evolution of the capacities of human self-organisation in natural history. According to present observational evidence in cosmic, planetary and biotic evolution, there must have occurred cataclysmic transitions in the degrees of organisational complexity of certain aggregations of matter that apparently elude any approach based on the notions of regularity and continuity as the conceptual cornerstones of intuitive as well as scientific evolutionism. To be sure, the formation of self-replicating entities (as the basic units of living systems) from prebiotic autocatalytic agents is familiar in organic chemistry.

"However, the specificity of physical forces on which the fidelity of self-replication is based, is limited. Improvements of fidelity could result only from catalytic support, where the catalyst... had to be reproducible, too. Translation of information inherited by the reproductive material became a requirement at this stage of evolution. The hurdle was immensely high... Required was a machine, but in order to produce it this very machine had to be available right away." (Eigen and Schuster 1979, p. IV)

Similar statements have been made by evolutionary biologists (e.g., Wilson 1975) concerning the natural history of the human cognitive and sociocultural capacities. The leap forward in mental evolution that separated *Homo sapiens* from his closest phylogenetic predecessors in an extremely short evolutionary time span "continues to defy self-analysis". Trends towards flexibility of social organisation in other primate species have been turned into a "protean ethnicity" in modern Man. Any phenotypic expression he may share with his phylogenetic relatives from some naive phenomenological perspective proves high-dimensional, culturally adjustable and almost endlessly subtle on closer inspection.

Observations like these have recently led to the resumption of doctrines postulated in epistemological debates on evolution earlier this century. They are known as the doctrines of *holism* and *emergence* (Elsasser 1975; Rosen 1978; Nagel 1979). Holistic ideas played a central role in the development of general system theory (Phillips 1976), and still govern the debates on sociobiology (Montagu 1980; Dawkins 1982; Rose et al. 1984) and evolutionary epistemology (Lorenz 1973; Popper and Eccles 1977; Riedl 1979; Sattler 1986). Re-examining the range of validity of holistic postulates in the life sciences proves useful in general, where problems of structural complexity are addressed, and, more specifically, in biobehavioural approaches to human sociocultural organisation.

We try explications of the notions of emergence and holism within the present conceptual framework, reconstructing the holistic terminology first. *Emergence* is the class of processes of hierarchical differentiation. (An explicit definition of the concept of process, together with examples, will be given in the next chapter.) Since increasing hierarchical differentiation of structures and systems implies increases in complexity, emergence is a subclass of evolution. For every n-level hierarchy of structures $\langle\langle M_1, S_1\rangle,\ldots,\langle M_n, S_n\rangle\rangle$ with $n \geq 2$, the rth stratum is called an *emergent structure relative to* the mth stratum if and only if $2 \leq r \leq n$ and $1 \leq m \leq r-1$ (cf. Rosen 1978, Sect. 4.5 and p. 112; Nagel 1979, Sect. 11. IV). Every R, $R \in S_r$, is called an *emergent attribute (property, relation) relative to* (the elements of) S_m if and only if $2 \leq r \leq n$ and $1 \leq m \leq r-1$.

We call the assertion "The whole is more than the sum of its parts" the *holistic principle*. Nagel (1979, Sect. 11.V) analysed this principle in various contexts where part-whole relationships arise and found it ambiguous in content mostly because of the intensional vagueness of the terms "whole", "part" and "sum". Considering the system-theoretic explications these terms have been given in the literature, one meets with similar ambiguities. Accordingly, some interpretations of the holistic principle must be discussed later in the particular contexts into which they belong. Here we concentrate on a familiar explication in terms of emergence saying,

"coupled systems (wholes) have emergent attributes relative to the attributes of their decoupled subsystems (parts)".

In order to give this postulate a (somewhat pedantically) correct reformulation within the terminology introduced thus far, we have to add one more definition to Section II.1.

For every s, s is a *coupled system* if and only if there is a two-level hierarchical structure $\langle\langle M_1, S_1\rangle, \langle M_2, S_2\rangle\rangle$ with the following properties: There are

1. a set V of systems with structure in $\langle M_1, S_1\rangle$;
2. a homogeneous relation C in V;
3. a positive integer k and systems $s_1,\ldots,s_k$ with structures in $\langle M_1, S_1\rangle$ and $\langle s, s_1,\ldots,s_k\rangle \in C$; so that
4. s is a coupling of $s_1,\ldots,s_k$ and $\langle M_2, S_2\rangle$ is a stratum appropriate to C.

Now the holistic principle can be phrased as

"coupled systems are defined in terms of structures (properties, relations) that are emergent relative to the defining structures of their decoupled subsystems".

Restatements of the holistic principle in terms of the concept of emergence have primarily been used to cope with the theoretical and epistemological problems posed by spontaneous evolutionary events. Scientists, system theorists and philosophers (e.g., Lorenz 1973, Chap. II; Elsasser 1975, Sect. 4.3; Popper 1977, Sect. P1.9) frequently conjectured that the evolutionary discontinuities characteristic of the origins of life and the human brain are effects of couplings between prebiotic, and between prehuman cerebral, systems respectively. For example, when molecules become arranged in such a way that they form cells and organisms, the appearance of qualitatively novel properties inherent in organic systems has been declared an evident effect of emergent evolution. However, such applications of the holistic principle yield little more than commonplace, namely that newly arising systems may have novel properties. In order to examine the adequacy of the concept of emergence to evolutionary theories, non-trivial explications of the holistic principle must be considered.

3 Holism Versus Reductionism

Empirical phenomena falling simultaneously in different research fields may constitute cross-disciplinary, intertheory relationships. This typically happens in the context of the universal evolution of matter. Intuitively, by the universal, or global, evolution of matter we mean the total of rearrangements of elementary particles into increasingly complex configurations of matter in cosmic and geological time. In a rough classification, these configurations may be schematically ordered as shown in Table II.1. The basic scientific disciplines concerned with intermediate stages in the global evolution of matter are listed together with systems of primary theoretical interest, and interactions between them.

The scheme of Table II.1 covers various hierarchies of structures and systems. For instance, such hierarchies result where the indicated interaction modes are coupling relations. Thus, biological structures may correspond to connexions

Table II.1. Sciences concerned with different levels of natural self-organisation. Similar schematic representations have been widely used in the literature to illustrate the global evolution of matter (cf. e.g., Popper 1977, pp. 16f)

```
┌─────────────────────────────────────────────────┐
│ Social and cultural sciences                    │
│                                                 │
│ Basic systems: Human individuals and groups     │
│                                                 │
│ Modes of interaction: Learning, tradition,      │
│ symbolic communication, syntactic language;     │
│ division of labour, economic exchange           │
└─────────────────────────────────────────────────┘
                        ▲
┌─────────────────────────────────────────────────┐
│ Life sciences                                   │
│                                                 │
│ Basic systems: Biomolecules, cells, organisms,  │
│ populations, ecosystems                         │
│                                                 │
│ Modes of interaction: Self-replication,         │
│ organismic reproduction; photosynthesis,        │
│ metabolism; selection, competition, predation;  │
│ animal communication, co-operation              │
└─────────────────────────────────────────────────┘
                        ▲
┌─────────────────────────────────────────────────┐
│ Physical sciences                               │
│                                                 │
│ Basic systems: Particles, atoms, molecules;     │
│ solid bodies, fluids, gases; charges, multipoles;│
│ radiation                                       │
│                                                 │
│ Modes of interaction: Fields and forces;        │
│ chemical reactions                              │
└─────────────────────────────────────────────────┘
```

Processes and stages of the universal evolution of matter (geological, organic, sociocultural) ▲

between physicochemical systems, and sociocultural structures to coupling relations between biological systems, with human individuals viewed as organisms. In the system–theoretical terminology introduced in Section II.1, these correspondences have been made precise in terms of the concepts of coupling degree and appropriateness of structures to coupling relations. Alternatively, using the holistic terminology, biological structures may be called emergent relative to the physicochemical structures of prebiotic systems, and sociocultural relations may be termed emergent attributes with respect to the biological attributes of human individuals.

As for the empirical hypotheses and theories available in the natural and social sciences, often no simple one-to-one correspondences exist between the ranges of applicability of these theories, on the one hand, and the levels of natural self-organisation, on the other. For instance, the laws of physics and chemistry clearly apply to cells, tissues and organisms to the extent to which living systems share the relevant physical properties with non-living entities. And despite its historically, culturally variable characteristics, *Homo sapiens* is no less an object of biological theories than any other animal species in natural history. This situation has led scientists and philosophers in the traditions of physicalism and biologism (biological determinism) to the following conjectures. Evolutionary changes in organic, and human cultural, structures and systems are nothing but effects of the physicochemical and biological, respectively, attributes of matter. To this empirical hypothesis corresponds a metatheoretic statement. It postulates that theories already exist, or can be found, for the physicochemical, biological and sociocultural levels of natural self-organisation, and an ordering exists for these theories such that the theories of more evolved organisational phenomena follow from those of more fundamental, evolutionarily antecedent structures.

The latter conjecture is sometimes called the *principle of reductionism* since it relates to a more general type of intertheory relationship known as the reducibility of one theory to another. In particular, the holistic doctrine of emergent evolution has been intended as the antithesis to reductionism and has been explicated, in this sense, as the thesis of the inherent irreducibility of (theories of) hierarchical organisation (Kanitscheider 1979). In order to examine this connotation of "emergence", we must first explain the relevant concepts.

3.1 Reduction of Theories

We concentrate on deterministic (i.e., non-probabilistic) and axiomatic theories. As for explicit representations, we restrict ourselves to elementary theories (having the logical syntax of a first-order predicate calculus with identity). These restrictions are not intended to resuscitate oversimplified views of empirical scientific theories of the kind the logico-empiricists might once have held. They are merely supposed to be adequate to the analysis of certain limited, though important, aspects of intertheory relations that are characteristic of theories of hierarchical organisation. The present mode of conceptualising scientific theories thus corresponds to the requirement mentioned in the *Preface*, namely to recast the frameworks of holism and reductionism into a form suited to reaching definite conclusions. One may reasonably doubt that because of their rich connotations, the previous verbal accounts of "interlevel theories" and "interfield connexions" (Wimsatt 1974, 1975; Maull 1977; Bechtel 1986) really come to grips with the intricacies of hierarchical organisation explained in Section II.1. In fact, the relevant conceptual problems, which have previously been pointed out by Causey (1977) at full depth, have apparently never been attempted to solve in a rigorous fashion.

The technical apparatus of the approach, of which we give a brief summary now, is explained in most elementary textbooks of formal logic and model theory (e.g., Ebbinghaus et al. 1984; Schwabhäuser 1971, 1972). The descriptive vocabularies of empirical scientific theories are finite sets of individual terms (constants, variables) and predicates (including operation symbols), which we designate respectively by lower case and capital Greek letters with indices where expedient. Capital Greek letters may also serve as sentential variables. We ambiguously use the sentential connectives "$\neg$", "$\Rightarrow$", "$\Leftrightarrow$", "$\wedge$", "$\vee$", the quantifiers "$\wedge$" (universal generalisation), "$\vee$" (existence), and the identity symbol "$=$" in the object and metalanguage. Expressions like "$\wedge \xi_1 \cdots \wedge \xi_m \cdots$" (or "$\vee \xi_1 \cdots \vee \xi_m \cdots$") may be written as "$\wedge \xi \cdots$" (or "$\vee \xi \cdots$"), while "$\xi$" may be shorthand for the list "$\xi_1, \ldots, \xi_m$" of individual terms ($m \geq 1$). The length of the abbreviated list will then be easy to identify from the context.

By a (*first-order*) *language* we understand the total of well-formed formulae (of the first order) the descriptive terms of which are drawn from some given vocabulary. Formulae, or, equivalently, *sentences* with no free variables are called *closed*, and *open* otherwise. Let L be a first-order language. If Φ is an m-place L-predicate, "ξ" is shorthand for "$\xi_1, \ldots, \xi_m$", and for every i with $i \leq m$, ξ_i is an individual L-term, then $\Phi(\xi)$ is an L-formula. If Ω is an L-formula, then the notation "$\Omega[\![\xi]\!]$" indicates that $\xi_1, \ldots, \xi_m$ occur free in Ω. For variables we stipulate that they do not occur free in $\Omega[\![\xi]\!]$ unless their symbols occur in the list abbreviated by "ξ".

Let M be a non-empty set. A *semantic interpretation I of L with the universe M* is a function which maps, as customary, the descriptive vocabulary of L into M and the relations defined in M (e.g., Ebbinghaus et al. 1984, Chap. III), while either $I(\Omega) = $ *true* or $I(\Omega) = $ *false* for every $\Omega \in L$. Thereby $I(\Omega)$ is recursively defined, in the usual manner, with reference to the number of logical constants occurring in Ω. If $I(\Omega) = $ true, I *satisfies* (is a *semantic model* of) Ω. For every subset $X \subset L$, I is a (*semantic*) *model* of X exactly if for every $\Omega \in X$, I satisfies Ω; and a sentence $\Psi \in L$ is said to *follow from X in L* (to be an *L-consequence* of X) exactly if every model of X is also a model of Ψ.

Let C_L be the function that assigns the set of L-consequences of X to every subset $X \subset L$. Then T is a *theory* (with the language L) if and only if $T = C_L(T)$. If T is a theory and $A \subset T$, then A is a set of *axioms* for T just in case $T = C_L(A)$. T is called an *empirical theory* under the semantic interpretation I if (a) $T \neq L$ (i.e., T is consistent), (b) I is a partial interpretation restricted to the descriptive constants and closed sentences of L, and (c) the universe M of I contains observable objects. (It is widely agreed among philosophers of science that the observability of objects and attributes and, hence, the empirical character of scientific theories depend on research contexts; cf. Hempel 1965; von Kutschera 1972). We do not explicitly mention the semantic interpretation of an empirical theory whenever specific empirical applications of this theory are irrelevant to the analysis. Let $T \subset L$, $T' \subset L' \subset L$, and $T = C_L(T)$; then T' is a *subtheory of T* with the language L' exactly if $T' = C_{L'}(T')$ and $T' \subset T$.

We are now in a position to explain what we mean by theory reduction and synthesis of theories. In science and philosophy there are numerous connotations of "theory reduction", ranging from replacement of inaccurate theories to intertheory relations involving the derivability of one theory from another (for discussion and recent applications, see Schaffner 1967; Wimsatt 1979; Balzer et al. 1984). We use a rather narrow concept of reduction which comes close to what the logico-empiricists originally intended. However, we do not want to give an exhaustive account of theory reduction here. We only insist that the present framework is suited to making a number of important points regarding the structure, dynamics and stability of hierarchically organised systems. In particular, we can analyse, in sufficiently exact terms, an intertheory relation which is in a sense complementary to theory reduction. Since we are exclusively interested in empirical scientific theories and, among these, in sufficiently idealised cases, we define the concept of reduction only for theories that are semantically interpreted and true under their respective interpretations.

Suppose that T_1, T_2 are elementary theories with the respective languages L_1, L_2, and let L be the language of $T_1 \bigcup T_2$. Then T_1 is said to *include T_2 on the basis of D* if and only if

1. for every descriptive constant of T_2 not occurring in T_1, there is an L-definition the definiens of which is an L_1-formula;
2. D is the total of these definitions; and
3. $T_2 \subset C_L(T_1 \bigcup D)$.

If these conditions are fullfilled and I_1 satisfies T_1, then, according to some well-known result of model theory (e.g., Schwabhäuser 1971, p. 117) there exists exactly one expansion of I_1 which satisfies $T_1 \bigcup D$. We denote this expanded semantic interpretation by "$\bar{I}_1^D$".

The theory T_2 under the semantic interpretation I_2 is *reducible to T_1* under the semantic interpretation I_1 if and only if

1. T_1 and T_2 are true under I_1 and I_2 respectively;
2. there is a set D so that T_1 includes T_2 on the basis of D; and
3. $I_2 = \bar{I}_1^D|_{T_2}$,
   where the symbol "$\bar{I}_1^D|_{T_2}$" denotes the restriction of $\bar{I}_1^D$ to (the descriptive terms and the sentences of) T_2.

If T_2 is reducible to T_1 under suitably chosen semantic interpretations, T_1 is said to *reduce T_2* under these interpretations. In particular, theories reduce their subtheories. Observe that the preceding definition can be viewed as a reconstruction of Nagel's (1979) classical concept of reducibility. Condition (2) amounts to Nagel's (1979, p. 354) conditions of "connectability" and "derivability", while (3) implies that D is a set of empirical truths (Nagel 1979, p. 355).

We also need the concept of the irreducibility of one theory to another, the meaning of which is obvious, however. We may therefore content ourselves with an example. For $i = 1, 2$, let $\langle M_i, S_i \rangle$ be a structure, and T_i an elementary theory, with the semantic interpretation I_i mapping the object constants of T_i into M_i, and the predicates of T_i into S_i. Let further $M_1 \bigcap M_2 = \emptyset$ so that $\langle M_1, S_1 \rangle$ and $\langle M_2, S_2 \rangle$ are incomparable (disparate). Then T_2 is irreducible to T_1 and vice versa. This result follows immediately from the disparity of the two structures. For if T_1 includes T_2 on the basis of D, the corresponding expansion $\bar{I}_1^D$ of I_1 still has the universe M_1. Hence, the condition (3) of reducibility cannot be satisfied.

3.2 Lower-Level Representations

We now proceed to examine two strong, or non-trivial, versions (A) and (B) of the holistic principle. Roughly speaking, if $\langle M, S \rangle$ is a structure, the weak, or trivial, explication of the holistic principle considered thus far asserts that every coupling of systems with structures in $\langle M, S \rangle$ induces an emergent structure relative to $\langle M, S \rangle$. In addition, the strong version (A) of holism asserts that, as an empirical postulate, the upper, or emergent, strata of hierarchical structures and systems are never causal, deterministic effects of the properties, relations and interactions of the corresponding lower levels of organisation. In order to make the terminology of this postulate precise, we must introduce some more definitions.

The notion of deterministic effect between structures and systems roughly means regular, recurrent interaction. In theoretical representations, interactions of this kind are usually described by cause-effect implications, or "causal laws", that is sentences of the form of generalised material implications, whereby the if-parts of the implications are no unreal conditional clauses. Throughout the present investigation, we use this type of sentence to characterise syntactically deterministic natural laws and similar generalisations of empirical hypotheses. We thereby treat the predicates "deterministic" and "causal" as synonyma although elsewhere there may be differences in meaning. The concept of causality frequently intends not only a deterministic cause-effect relationship between, but also a time-ordering of, events. The time dependence of events, however, is generally not made explicit in our analyses. Accordingly, for every two-level hierarchical structures $\langle \langle M_1, S_1 \rangle, \langle M_2, S_2 \rangle \rangle$, a deterministic effect between $\langle M_1, S_1 \rangle$ and $\langle M_2, S_2 \rangle$ is any regular interlevel dependence of facts and events. We explicitly define this type of interaction only for effects of $\langle M_1, S_1 \rangle$ on $\langle M_2, S_2 \rangle$. Reverse effects can then be characterised in an analogous fashion.

For the sake of lucidity, we first introduce a suitable terminology and notation. We consider set-theoretic formulae of the first order, meaning formulae that contain no non-descriptive constants other than the symbols "$\in$", "$=$", and the familiar sentential connectives and quantifiers of a first-order logical calculus. However, if $\mathscr{F}$ is a set-theoretic formula of the first order, relation variables are admitted in $\mathscr{F}$, provided they occur free, Clearly the symbols "$\cup$", "$\subset$", "$\times$", etc., are also admitted in $\mathscr{F}$. In the metalanguage, we use boldface letter symbols to designate set-theoretic constants, variables and sentences (with the exception of non-descriptive constants). So "**x**" designates "x" and "$\vee z(z \in A \cap B)$" is a metatheoretic name of the formula "$\vee z(z \in A \cap B)$", etc. We use the expressions "$\mathscr{F}[\![X_1, \ldots, X_n]\!]$" to indicate that the relation variables "X_1", ..., X_n", but no other (object or relation) variables, occur free in $\mathscr{F}$. When $\langle M, S \rangle$ is a structure, $\mathscr{F}$ is said to be *based on* $\langle M, S \rangle$ if and only if $a \in M$ and $R \in S$ whenever the object constant "a" and the relation constant "R" occur in $\mathscr{F}$. The formula $\mathscr{F}$ is said to *state a fact about* (or *factual situation in*) $\langle M, S \rangle$ exactly if $\langle M, S \rangle$ is a structure, $\mathscr{F}$ is based on $\langle M, S \rangle$ but contains no relation variables at all and no free-object variables, and $\mathscr{F}$ is true in $\langle M, S \rangle$.

Suppose $\langle \langle M_1, S_1 \rangle, \langle M_2, S_2 \rangle \rangle$ is a two-level hierarchical structure and, without loss in generality, $M_2 = \mathscr{P}(M_1)$. The latter assumption allows one to use object variables ranging over M_2 as set variables for arbitrary subsets of M_1. The stratum $\langle M_2, S_2 \rangle$ *has a representation in* $\langle M_1, S_1 \rangle$ (casually: a *lower-level representation*) if and only if the following condition holds: If Q is an n-place relation with $Q \in S_2$ and $n \geq 1$, there exists a formula $\mathscr{F}$ based on $\langle M_1, S_1 \rangle$, with n set variables, for which

$$\wedge X_1 \cdots \wedge X_n(\langle X_1, \ldots, X_n \rangle \in Q \Rightarrow \mathscr{F}[\![X_1, \ldots, X_n]\!]) \tag{II.15}$$

holds.

This concept of representing $\langle M_2, S_2 \rangle$ in $\langle M_1, S_1 \rangle$ specifies a type of cross-level correspondence between factual situations in hierarchies. Roughly, $\langle M_2, S_2 \rangle$ having a representation in $\langle M_1, S_1 \rangle$ means that each relation contained in S_2 correlates with a class of lower-level factual situations according to (II.15). As an example, consider Eqs. (II.12) and (II.13) for $r = 2$, with the two-level hierarchy $\langle \langle M_1, S_1^* \rangle, \langle M_2, S_2 \rangle \rangle$. For every $Q \in S_2$ there exists a relation W, $W \in S_1^*$, so that

$$\wedge q(q \in Q \Rightarrow [\![q]\!] \subset W), \tag{II.15'}$$

where the formula "$[\![q]\!] \subset W$" is based on the lower stratum $\langle M_1, S_1^* \rangle$.

3.3 Cross-Level Deterministic Effects

The stratum $\langle M_2, S_2 \rangle$ is said to be a *partial deterministic effect of*, or, equivalently, *partially determined by* the basic stratum $\langle M_1, S_1 \rangle$ if and only if

1. for every $Q \in S_2$ there is a homogeneous k-place relation C in S_1 ($k \in \mathbb{P}$) so that C is transformed into Q under (II.6) and (II.7), and
2. there is a formula $\mathscr{F}$ based on $\langle M_1, S_1 \rangle$, with k relation variables "R_1",…,"R_k", for which it holds that

$$\wedge R_1 \cdots \wedge R_k(\mathscr{F}[\![R_1, \ldots, R_k]\!] \Rightarrow \langle R_1, \ldots, R_k \rangle \in C). \tag{II.16}$$

The structure $\langle M_2, S_2 \rangle$ is *completely determined by* $\langle M_1, S_1 \rangle$ if it is partially determined by $\langle M_1, S_1 \rangle$ and

$$\wedge R_1 \cdots \wedge R_k(\mathscr{F}[\![R_1, \ldots, R_k]\!] \Leftrightarrow \langle R_1, \ldots, R_k \rangle \in C) \tag{II.17}$$

holds in addition to (II.16). When $\langle M_2, S_2 \rangle$ is completely determined by $\langle M_1, S_1 \rangle$, it has a trivial representation in $\langle M_1, S_1 \rangle$ by virtue of (II.17). We then call this representation a *deterministic*, or *causal*, *lower-level representation* because according to

$$C = \{\langle R_1, \ldots, R_k \rangle \mid \mathscr{F}[\![R_1, \ldots, R_k]\!]\}, \tag{II.17'}$$

each relation $Q \in S_2$ is completely determined by that class of lower-level factual situations by which it is represented. In fact, in physical, biological and sociological applications the formulae (II.15) to (II.17) will obey suitably chosen empirical deterministic laws. In order to illustrate these notations and definitions, we refer to the macrosystems mentioned in Section II.1. These systems, and their structures, clearly have lower-level causal representations in their respective microstructures.

For every pair $\langle \langle M_1, S_1 \rangle, \langle M_2, S_2 \rangle \rangle$ of structures, $\langle M_2, S_2 \rangle$ is said to be *emergent relative to* $\langle M_1, S_1 \rangle$ *in the strong sense* if and only if

1. $\langle \langle M_1, S_1 \rangle, \langle M_2, S_2 \rangle \rangle$ is a two-level hierarchical structure (condition of emergence in the weak sense), and
2. $\langle M_2, S_2 \rangle$ has no representation in $\langle M_1, S_1 \rangle$.

In particular, (2) excludes the possibility of $\langle M_2, S_2 \rangle$ having any lower-level causal representation.

3.4 Holism Reconsidered

Now the strong version (A) of holism consists in the empirical postulate that for every two-level hierarchical structure $\langle \langle M_1, S_1 \rangle, \langle M_2, S_2 \rangle \rangle$ in the inanimate and living world, $\langle M_2, S_2 \rangle$ is emergent relative to $\langle M_1, S_1 \rangle$ in the strong sense. As a corollary, (A) states that in hierarchies of physical, organic and sociocultural structures and systems the upper strata do not depend exclusively and completely upon lower-level forms of organisation. It is straightforward to extend these definitions and assertions to n-level hierarchies with $n > 2$.

Since (A) denies higher strata of natural self-organisation to admit lower-level representations, (A) is sometimes said to state the "irreducibility of hierarchical organisation" (see Nagel 1979, pp. 367-374) or "irreducibility of hierarchical complexity" (Elsasser 1975, pp. 94, 108-112). However, we do not adopt this terminology because we reserve the concept of reducibility and related notions for intertheory relationships.

Like the principle of reductionism, the non-trivial postulate of emergence has also been phrased metatheoretically, namely as the principle (B) of the irreducibility of deterministic theories of hierarchical organisation and emergent structures (cf. Nagel 1979, pp. 366-380). Stated for elementary theories, it reads:

"Given any two-level hierarchical structure $\langle \langle M_1, S_1 \rangle, \langle M_2, S_2 \rangle \rangle$ and theories T_1, T_2 under respective semantic interpretations I_1 and I_2 relative to $\langle M_1, S_1 \rangle$ and $\langle M_2, S_2 \rangle$. Let T_1 and T_2 be true under I_1 and I_2 respectively. Then T_2 under I_2 is irreducible to T_1 under I_1."

Thereby, the phrase "theory T under semantic interpretation I relative to structure $\langle M, S \rangle$" simply states that I maps the individual T-constants into M, and the T-predicates into S.

A few points can readily be made on the strong versions of holism. As a general empirical principle, (A) is false. There are numerous counterexamples to the doctrine that the upper strata in hierarchies of structures are emergent in the strong sense. The macrosystems of Section II.1 and their microstructures constitute such counterexamples. The flaw in this version of holism in the strong sense is ignorance of the fact that the structures and hierarchies considered in science and everyday life are selected by scientists and non-scientists from an extreme multitude of structures in the real world. And frequently it is possible and desirable to select the structures of one's interest precisely in such a manner that no stratum is emergent in the strong sense relative to any other stratum under consideration.

As a metatheoretic theorem, the principle (B) of emergence holds for elementary theories. We have seen above that the elementary theories T_1 and T_2 are irreducible to one another when they are semantically interpreted relative to incomparable structures. In fact, if the structures $\langle M_1, S_1 \rangle$ and $\langle M_2, S_2 \rangle$ form a hierarchy, they are trivially incomparable. However, this instance of irreducibility bears no general significance as the irreducibility may vanish if the restriction of T_1 and T_2 to elementary theories is relaxed and the semantic structures of T_1 and T_2 are appropriately modified. Such a modification would have to be devised so as to satisfy an analogue of condition (2) of reducibility. One would have to (a) augment the descriptive vocabulary of T_1 by a set Δ of second-order predicates, (b) extend the domain of I_1 into Δ, and the range of I_1 into S_2, (c) redefine the T_2-constants in terms of the constants of T_1 and the elements of Δ, and (d) derive T_2 from T_1 and the definitions subsumed under (c).

This procedure may indeed prove successful if $\langle M_1, S_1 \rangle$ has a deterministic effect on $\langle M_2, S_2 \rangle$. However, we do not explore the conditions here under which reductions among non-elementary theories are in fact feasible. We simply refer to the examples extensively treated in the literature. A case in point is the derivation of the laws of rigid-body mechanics from those of particle mechanics (Adams 1959; Pearce 1982). It not only exhibits the aspect of hierarchical stratification, with rigid bodies viewed as compound systems of particles, but also cross-level deterministic effects. For the motion of the rigid bodies is completely described, according to the classical laws, by the motion of the corresponding constituent particles, and the forces acting on rigid bodies are similarly obtained from an appropriate composition of the components of force acting on the particles. Thus holists are misled when inferring (B), as a general principle, from special instances of irreducibility arising in a very restricted class of theories with incomparable semantic structures. Moreover, we shall argue that, considering a serious paradox, the theories in this class do not even admit evolutionary applications. More precisely, we shall argue that there is no (consistent theory of) emergent evolution in the strong sense of holism.

Finally, we briefly and informally mention the possibility that in hierarchies of structures and systems the upper strata have deterministic effects on the lower ones. Such effects, which have been subsumed under the concept of downward causation by Sperry (1965), Campbell (1974) and others, now underlie much of the holistic reasoning in the (neuro-)physiologically oriented behavioural sciences. The proponents of holism maintain that in hierarchical organisations with effects of downward causation, the lower-level attributes and interactions not only do not determine, but are themselves effects of, the upper strata. Holists are thus led to the conclusion that in hierarchies exhibiting this kind of cross-level relation, no particular stratum (or part) controls the entire hierarchical structure (i.e., the hierarchy "as a whole") in an exclusive and complete fashion.

The problem with this application of holism is once more precipitate generalisation – apart from the question of why some holists deny deterministic effects directed up the hierarchy while admitting downward causation. The answer to the question of what determines what in stratified systems not only varies with the types of interactions, and hierarchies of interactions, known to natural and social scientists. It is also contingent upon each particular theory applied to hierarchical organisations. Counterexamples to the holistic doctrine, as applied to downward causation, abound in cybernetic approaches to cross-level relations. (By cybernetics we mean the total of scientific disciplines concerned with feedback interactions between systems.) A case in point is the self-regulation of animal social behaviour through natural selection. According to the combined game-theoretical, cybernetic account of Darwinian natural selection, which has become known as evolutionary game dynamics (Taylor and Jonker 1978; Hofbauer and Sigmund 1984), the social structure of an animal group is determined exclusively by the behavioural attributes of the individual group members, provided the group's ecological and social environment remains stable. Social groups thus form two-level stratified systems of behavioural stimuli (inputs) and responses (outputs). The group structure, on its part, determines the evolution of the individuals' behaviour patterns via selection. In this sense, behavioural differences between individual organisms are causes and effects of the selection process, which is exactly what the term "self-regulation of social behaviour through natural selection" means. The many approximations and simplifying assumptions inherent in the game-dynamical approach to sociobiology clearly do not impair the relevance of this example to downward causation.

3.5 Informal Statement of Results

The principles of emergence and holism have been reconstructed in logical and semantic terms and then analysed with reference to hierarchical structures. For the purposes of this analysis, several familiar concepts had to be redefined explicitly (i.e., made sufficiently precise). By a *theory* we mean the total of sentences of a given language that follow in this language from a given set of axioms. *Empirical theories* are semantically interpreted relative to real-world structures. A theory T_1 has been said to *reduce* another theory T_2 if all the T_2-concepts can be defined in terms of T_1 such that these definitions preserve the empirical truths of T_2 and, together with T_1, imply T_2. This concept of theory reduction is rather narrow and does not encompass all intertheory relationships of reducibility known to scientists and philosophers. But nonetheless, it has many realistic applications and proves particularly suitable for the logical and semantical analyses at which we aim.

Using these notions of empirical, scientific theory and reducibility of theories, we have reconstructed the principles of reductionism and holism both theoretically (i.e., as empirical postulates) and metatheoretically. As an empirical postulate, the principle of reductionism roughly states that the evolution of organic and sociocultural structures is nothing but an outcome of the physicochemical attributes of matter. Metatheoretically, reductionism implies that theories can be established, or already exist, for the physicochemical, biological and sociocultural levels of natural self-organisation such that the theories of more fundamental, less complex structures reduce those of more evolved organisational phenomena. Holism, on its part, has been characterised as the antithesis to reductionism. The weak, or trivial, version of holism mentioned at the end of the preceding section implies that hierarchical evolution gives rise to previously non-existing ("emergent") structures and systems with novel attributes. Now the principle of holism, or emergence, in the strong (non-trivial) sense states that – as an empirical principle – the upper strata of hierarchical structures are never causal, deterministic effects of the properties, relations and interactions of the corresponding lower levels of organisation. After having given the notion of cross-level deterministic effect an adequate and sufficiently precise meaning, we arrived at the conclusion that the empirical principle of holism in the strong sense is generally invalid. Counterexamples to this principle can be found among the micro- and macrosystems and their associate hierarchical structures discussed in Section II.1. The fundamental flaw in this version of holism in the strong sense is ignorance of the fact that the structures and hierarchies considered in science and everyday life are selected by scientists and non-scientists from an extreme multitude of perceivable structures in the real world. And frequently it is possible and desirable to select the structures of one's interest precisely in such a manner that none is emergent in the strong sense relative to any other structure under consideration.

When phrased metatheoretically, the principle of holism in the strong sense proposes the irreducibility of theories of hierarchical organisation. It has been shown that this proposition holds for elementary theories that are semantically interpreted relative to different levels of organisation in multilevel hierarchies. But it is incorrect to infer this irreducibility as a general principle from a very restricted class of theories with the logical structures of first-order predicate calculi. This irreducibility may indeed vanish if non-elementary theories are considered.

4 The Paradox of Emergent Evolution

This section deals with the inadequacy of the concept of emergence to theories of evolutionary processes. We show that the principle (A) of holism in the strong sense introduces a paradox into evolutionary theories of hierarchical organisation. The paradox arises from the fact that the class of emergent evolutionary processes is empty. By emergent evolution holists mean the total of processes through which the higher strata of hierarchies arise from the lower ones so that – according to (A) – the former are emergent relative to the latter in the strong sense. For details and historical notes on the notion of emergent evolution we again refer to Nagel's book (1979, pp. 374-380).

By an *evolutionary theory* we mean any theory which, firstly, contains the two-place predicate "evolves from" defined for structures and systems, and from which, secondly, sentences follow of the kind "Σ_2 evolves (evolved, etc.) from Σ_1". The symbols "Σ_1", "Σ_2" may denote either structures or systems. The precise meaning of "Σ_2 evolves from Σ_1" will be explained in Section III.4 together with the definition of the concept of evolutionary process. For the present it is sufficient to note that "Σ_2 evolves from Σ_1" implies the sentence "Σ_2 is (structurally/behaviourally) more complex than Σ_1", corresponding to the concept of evolution presently limited to increases in structural/behavioural complexity. Eventually, by a *theory of emergent evolution* we mean any evolutionary theory which contains the principle (A) of holism in the strong sense as an empirical axiom.

A few remarks may be useful at this point. Recall that we do not exclude the possibility of defining the concept of evolution in alternative ways, for instance, as the total of processes of natural selection and adaptation. Thus other notions of evolutionary theory may exist in which the assumed implication is generally invalid. Nor do we deny the existence of transitions and processes other than evolutionary ones connecting structures or systems. However, such non-evolutionary events, and their theories, are simply not the topic of the present investigation. We also ignore the problem that in evolutionary theories the expression "evolves from" must be a higher-order predicate. This is the case if, for instance, T is a non-elementary theory, "Σ_2 evolved from Σ_1" is a sentence of T, and Σ_1, Σ_2, respectively, are two- and three-level hierarchical structures. In Section III.3 we rather show how to treat hierarchies of structures and systems within the logical and semantic frameworks of elementary theories.

4.1 Emergent Evolution: A Crude Misconception

Let T be now an arbitrary evolutionary theory proposing that Σ_2 evolved from Σ_1, where Σ_1 and Σ_2 are structures. Then, "Σ_2 is more complex than Σ_1", and *a fortiori*,

"Σ_1 and Σ_2 are commensurable in complexity", are sentences of T. Furthermore, suppose that Σ_2 is emergent relative to Σ_1 in the strong sense. For example, Σ_1 and Σ_2 may form the two-level hierarchical structure $\langle \Sigma_1, \Sigma_2 \rangle$. It follows immediately that they are disparate and, hence, incommensurable in complexity. One thus gets the paradoxical result that every evolutionary theory of emergent structures is inconsistent, and no structure can be said to evolve from lower levels of organisation relative to which it is emergent in the strong sense of holism.

The paradox of emergent evolution clearly arises from the application of the predicates "evolves from" and "is more evolved (complex) than" to structures which are by definition incommensurable in complexity. In the context of (II.12) to (II.14) we have already indicated the way in which the holistic paradox can be avoided. It has been shown that every multilevel hierarchy with basic stratum Σ_1 can be transformed into a uniquely determined structure Σ_1^* which is commensurable in complexity with Σ_1. On the basis of this transformation, namely (II.12), the evolutionary relationship between the strata Σ_1, Σ_2 of the two-level hierarchy $\langle \Sigma_1, \Sigma_2 \rangle$ can be correctly restated as follows: "The two-level hierarchy $\langle \Sigma_1, \Sigma_2 \rangle$ evolved from (is more complex than, etc.) the one-level hierarchy Σ_1."

The latter formulation is not merely terminological hair-splitting; it rather expresses a strictly non-holistic view of hierarchical organisation. In Section III.3 we show that the total of facts about Σ_2 can be mapped into the set of facts about Σ_1^*, where Σ_1^* is the transformed structure of $\langle \Sigma_1, \Sigma_2 \rangle$ according to (II.12) and (II.13). The upper stratum Σ_2 is thus not only mapped into a structure that is commensurable in complexity with the basic stratum Σ_1. We also show that Σ_2 can be given a lower-level representation exactly of the kind which the strong version (A) of holism denies to exist. That such lower-level representations do exist clearly does not mean that they are always useful or easy to construe in realistic cases. Needless to say that it may even be *practically* impossible to give explicit microscopic representations to all the macrosystems with which scientists are concerned. But it makes a difference whether one takes holism as a pragmatic attitude, or accepts the holistic principle as a universal empirical law.

On the other hand, the present view of evolved hierarchical stratification is, though non-holistic, not necessarily a reductionistic one. From "Σ_2 admits a lower-level representation in Σ_1^*" by no means follows that for every theory T_2, which is true under the semantic interpretation I_2 relative to Σ_2, there exists a theory T_1 and a semantic interpretation I_1 for T_1 relative to Σ_1^* so that T_2 under I_2 is reducible to T_1 under I_1. Nor is every lower-level representation of Σ_2 a deterministic one, as is evident from the relevant definitions.

4.2 A Note on Stochastic Evolutionary Theories

For completeness, another type of holistic reasoning in natural science must be briefly mentioned here. It eludes the preceding analysis insofar as it refers to stochastic events, non-causal empirical laws and probabilistic theories. This kind of holism, too, has been applied to the evolutionary, epistemological problems posed by the origins of life and the human brain. Proponents of holism as applied to stochastic evolutionary events are Monod (1970), Elsasser (1975) and Popper (1977), to name a few. Their notion of emergent evolution can be summarised as follows:

The Paradox of Emergent Evolution

In physics and the life sciences, complex stratified systems can in principle be represented and theoretically described in terms of the basic structures on which the hierarchies are respectively built. For example, the physiological effects of a protein molecule are determined by the arrangement of the constituting amino acids along the protein chain. However, in a vast variety of cases the relevant microscopic (i.e., lower-level) arrangements and interactions require probabilistic laws and theories for their description. Based on the combinatorial principles of statistical physics and equilibrium thermodynamics, many theories assign extremely small, though non-zero, probabilities to certain events characteristic of hierarchical differentiation. As for the example of protein structure, a very short protein chain may consist of about 100 amino acids drawn from the 20 well-known standard types of amino acid. One may then ask for the reaction rate at which, under physically realistic conditions, protein molecules with some specified spatial configuration synthesise in a medium of arbitrarily arranged amino acids. The reaction probability has been estimated at much less than 10^{-122} per protein molecule per second (Eigen 1971; Nicolis and Prigogine 1977, p. 23; cf. also Küppers 1983, 1984). Accordingly, at its present age of about 10^{17} s the physical universe seems far too young to make the spontaneous formation even of a very primitive biochemical substratum a realistic event.

These considerations and numerical estimates have motivated definitions and assertions of the following kind. Let $\langle \Sigma_1, \Sigma_2 \rangle$ be a two-level hierarchical structure, and let T be a theory which assigns the probability $p \neq 0$ to the event that $\langle \Sigma_1, \Sigma_2 \rangle$ evolves from Σ_1. Then Σ_2 is called *emergent relative to* Σ_1 if and only if p is numerically of the order of 10^{-n}, where n is a large number. (In his slightly different terminology, Elsasser (1975) uses the term "autonomous" instead of "emergent".) The principle of emergent evolution now roughly reads as follows: In the prebiotic, organic and sociocultural structure of matter, numerous strata $\Sigma_1, \Sigma_2, \ldots$ can be distinguished such that for $k \geq 2$ Σ_k is emergent relative to Σ_{k-1} in those theories that respectively describe $\langle \Sigma_{k-1}, \Sigma_k \rangle$.

We do not intend to examine in detail these definitions and principles of emergence, which task leads far beyond the scope of the present investigation (for references and discussions, see Küppers 1984). We rather close this section with two programmatic remarks. Firstly, if the global evoluion of matter were emergent in the sense of being governed by extremely improbable events, no evolutionary laws and theories could be empirically true other than those that assert the evolutionary processes to be virtually impossible. It is thus absurd to postulate theories of emergent evolution. Secondly, the holists seem to underrate the narrow restriction of their principles to the theories available for the time being. On further research, spatiotemporally coherent macrostructures may well prove much less "autonomous" than the combinatorics of microphysical or microchemical reaction probabilities would suggest. Recent development in statistical physics, biochemistry and neuroscience apparently supports this expectation (Eigen and Winkler 1975; Nicolis and Prigogine 1977; Haken 1978, 1983; Eigen and Schuster 1979; Eigen 1983; Babloyantz 1986).

III The Concept of Unified Theory

1 Synthesis Versus Reduction of Theories

Considering the spectrum of interpretations of the holistic principle, this principle proves either trivial or false. However, since evolutionary debates in science and philosophy continue hinging on notions of holism and emergence (Rose et al. 1984; O'Neill et al. 1986; Sattler 1986), it is hard to embark on an analysis of current evolutionary concepts and theories without getting entangled in the holism-reductionism controversy. The holistic objections against ethological and sociobiological accounts of human social relations povide examples that are typical of this situation. When holists dismiss biological explanations of sociocultural phenomena without further examination, they allude to the "reductionist" premises of these explanations, that is, insensitivity to the allegedly emergent character of human culture. Whatever the merits or limits of biological contributions to social science may be, the holists ignore the problem of making biological and cultural levels of organisation comparable and, above all, commensurable in complexity in order to put sociocultural structures into an evolutionary perspective. The application of concepts of emergence to evolutionary problems in social science is thus illustrative in quite a general sense. It shows that holism is particularly well designed to obscure, rather than explain, evolutionary processes.

On the other hand, the holism-reductionism dichotomy does not exhaust the spectrum of the relevant intertheory relations. In science one often meets with theories which simultaneously reduce, or even contain as subtheories, several other theories different in range of applicability. The former theories are known as "unified (unifying, synthetic) theories". Sometimes they are also said to "synthesise" the other theories. Problems of a unified field theory govern the foundations of physics, and 20th century evolutionary biology has produced a variety of syntheses (cf. Bechtel 1986) such as neo-Darwinian "synthetic theory" (Huxley 1942), sociobiological theories (Wilson 1975) and unified theories of natural selection (Dawkins 1982; Milkman 1982; Schuster and Sigmund 1983). In the following sections we examine the usefulness of unified theories especially for the treatment of hierarchically stratfied, evolving systems.

The notions of unified theory customary in science can be reconstructed in different ways, depending on the respective concepts of theory and theory reduction used for explicit definition. The definition we adopt here is much in the spirit of Causey's (1977, Chap. 6) book and fits into the conceptual framework of the previous chapter. For $n \geq 2$, let $T_1, \ldots, T_n$ be theories with corresponding semantic interpretations $I_1, \ldots, I_n$. Suppose that for every pair, i, j of positive integers with $i \leq n, j \leq n$ and $i \neq j$, T_i does not logically imply T_j (condition of independence of $T_1, \ldots, T_n$). Then the theory T is a *unified theory with respect to* $T_1, \ldots, T_n$ exactly if for every

i, $i \leq n$, T under the interpretation I reduces T_i under the interpretation I_i, and $T \neq T_i$. The latter condition – like the condition of independence of $T_1, \ldots, T_n$ – excludes trivial cases of unified theories. If T is a unified theory with respect to $T_1, \ldots, T_n$ we also call the theories $T_1, \ldots, T_n$ *simultaneously reducible to* T, and T is said to *synthesise* $T_1, \ldots, T_n$.

The significance of syntheses, as opposed to reductions, of theories can be exemplified best in terms of the theoretical problems raised by the hierarchical organisation of matter. Assume that for some integer $n \geq 2$, $\Sigma = \langle \langle M_1, S_1 \rangle, \ldots, \langle M_n, S_n \rangle \rangle$ is an n-level hierarchical structure. As pointed out above, there has been considerable interest in the question of whether there exist theories $T_1, \ldots, T_n$ so that for every i, $1 \leq i \leq n$, T_i is semantically interpretable relative to $\langle M_i, S_i \rangle$ and for every j, $2 \leq j \leq n$, T_j under I_j is reducible to T_{j-1} under I_{j-1}. For example, let $n = 4$ and T_4 be the theory of Mendelian genetics (with Mendel's original two laws as empirical axioms), T_3 a theory of biomolecular genetics, T_2 a theory of prebiotic chemistry, and T_1 a theory of atomic physics as applied to chemical bonds. The structures described by these theories are hierarchically ranked. Mendelian genes are basically arrays of nucleotides (segments of DNA-polynucleotides), each nucleotide consists of prebiotic molecules (nitrogenous bases, sugar, phosphate) which themselves do not polymerise, and so forth. Accordingly, there is vast literature on the questions of whether (the theory of) Mendelian inheritance can be explained by theories of biomolecular genetics, whether the theories of the biochemical apparatus of the cell can be reduced to theories of molecular bonding, etc.(Sober 1984, Part VI).

Alternatively, there may be a unified theory with respect to $T_1, \ldots, T_n$ rendering pointless any dispute about the successive reducibility of $T_n, T_{n-1}, \ldots$ to T_1. This case arises if T is a theory with a descriptive language and logical syntax of the nth order so that T is interpretable relative to Σ and reduces $T_1, \ldots, T_n$ under $I_1, \ldots, I_n$, simultaneously. The results and examples of the following sections suggest that constructing unified theories is indeed a promising and frequently rewarding research strategy for evolutionary approaches to multilevel systems. This conclusion largely agrees with the views previously taken by philosophers of science such as Wimsatt (1974, 1976). In Sections III.2 and III.4 we outline a scheme of how to unify a given sequence of theories. This scheme applies to theoretical representations of parameter families of systems, hierarchically stratified and others. In fact, the results of Section III.3 imply that within the limits of the present approach first-order theories suffice for unified theoretical descriptions of families of systems of either kind.

The difference between successive and simultaneous reduction of $T_1, \ldots, T_n$ is schematically depicted in Fig. III.1. The figure illustrates the fact that reducibility as intended by conventional scientific reductionism is something quite different from simultaneous reducibility in the sense of unified theories. Since the trivial case has been excluded in which a theory T_2 reducing another theory T_1 synthesises T_1 and T_2, a synthesis of T_1 and T_2 requires one more theory whose empirical content will generally be more inclusive, and whose formal structure will be richer than the content and structure of the joint theory $C_L(T_1 \cup T_2)$ with the language L of $T_1 \cup T_2$. For example, neo-Darwinian synthetic theory, which has been considered elsewhere (Geiger 1988a, b) relativises and, in a sense, corrects the theories of Mendelian inheritance and Darwinian selection, and makes them logically consistent.

It goes without saying that the present concept of unified theory intends an idealised type of intertheory relation as compared to what is usually called a unified

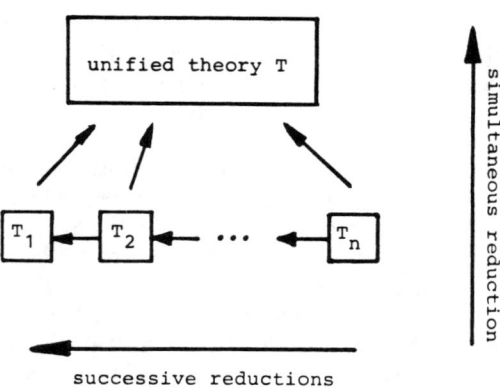

Fig. III.1. Schematic representation of intertheory relations. Successive reductions of the theories $T_n, T_{n-1}, \ldots$ to T_1 as compared to simultaneous reduction of $T_1, \ldots, T_n$ to the unifying theory T. For the sake of expedience, the semantic interpretations of the theories are not made explicit in this scheme. For a similar conception of unified theory, see Causey (1977)

theory in the empirical sciences. This discrepancy, however, is irrelevant to the scope of the present investigation.

2 Synthesis of Theories Through Parametrisation of Laws

The type of unified theory introduced in this section has numerous applications in the empirical sciences. Although parametrisation of laws is based on a simple logical principle, in practice it will turn out to be a rather non-trivial matter. Again we restrict our explicit definitions of the relevant concepts to first-order languages and theories.

Assume that T is an empirical theory under some suitably chosen semantic interpretation of its descriptive language. Then the closed T-sentence Ω is an *empirical law* (*generalised empirical hypothesis*) exactly if (a) Ω is not logically equivalent to any T-sentence containing no quantifiers and (b) Ω is non-analytic in T, meaning that neither Ω nor $\neg \Omega$ follows from the explicit definitions in T. Empirical laws with an initial universal quantifier are said to have *normal form*. The empirical law Ω is called an *empirical axiom*, or *postulate, of T* if $\Omega \in A$, where A is a set of axioms for T. Here we restrict attention to theories that follow from finite sets of empirical axioms.

2.1 Parametrisation of Laws

Suppose T is a first-order theory with the language L, and T is consistent. For $n \geq 2$, let $\Lambda, \Omega_1, \ldots, \Omega_n$ be L-sentences of the syntactic structure of empirical laws in normal form. In other words, there are positive integers $m, m_1, \ldots, m_n$ and L-formulae $\Psi[\![\zeta]\!]$ and $\Phi_i[\![\xi_i]\!]$ in which "ζ" and "ξ_i" abbreviate lists of L-variables of respective length m and m_i, so that

$$\Lambda = \wedge \zeta \Psi[\![\zeta]\!]. \tag{III.1a}$$

$$\Omega_i = \wedge \xi_i \Phi_i[\![\xi_i]\!], \quad i \leq n. \tag{III.1b}$$

Then Λ is said to *parametrise* $\Omega_1,\ldots,\Omega_n$ in T if and only if

1. $\Lambda \in T$;
2. $m_i < m$;
   and
3. there are $m - m_i$ individual constants – the list of which can be abbreviated by "η_i" – so that

$$\wedge \xi_i(\Psi[\![\eta_i, \xi_i]\!] \Rightarrow \Phi_i[\![\xi_i]\!]), \quad i \leq n, \tag{III.2}$$

is a sentence of T.

Evidently, if Λ parametrises $\Omega_1,\ldots,\Omega_n$ in T, then $\Omega_i \in T$ for every $i \leq n$.

The meaning of this definition can be intuitively circumscribed as follows. Consider the case $m = 2$ with the variables λ and ξ, and the constant η. Assume that η does not occur in $\Psi[\![\lambda, \xi]\!]$. Now one often wants to study laws "of the form" $\wedge \xi \Psi[\![\eta, \xi]\!]$ rather than $\wedge \xi \Psi[\![\eta, \xi]\!]$ itself, meaning that the value of η is of no particular interest to the inquiry. One may then consider the formula $\wedge \xi \Psi[\![\lambda, \xi]\!]$ in which case λ is usually referred to as a "parameter". Here, we call the pair λ, ξ a *partition of ζ into the state parameter λ and system variable ξ*, with an obvious generalisation to the case $m \geq 2$ in which "λ", "ξ" and "ζ" are shorthand for lists of L-variables of respective length $m', m - m'$ and m ($1 \leq m' < m$). For every such partition, $\wedge \xi \Psi[\![\lambda, \xi]\!]$ is equivalent to an L-sentence $\Delta(\lambda)$ with m' free variables corresponding to λ and associate (compound) L-predicate Δ. If I is a semantic interpretation of T, the set $I(\Delta)$ is termed the *state set*, or *state space*, of the partition. The reason for adopting this terminology is that laws of the type (III.1) will be used below to describe state-dependent behaviours of systems, with the meaning of "state" as explained in Section II.1. The arbitrariness in partitioning a given list of variables into sequences λ and ξ of optional length reflects the fact that for interacting objects in the real world the state view and the system view depend on the observer's contingent conceptual distinction. In (III.2) η_i is thus not required to have the same number of components for each $i \leq n$. However, for every $i \leq n$, η_i uniquely determines a partition of ζ into state and system variables with corresponding predicate Δ_i so that the *global state space P* can be defined as

$$P = \bigcup_{i=1}^{n} I(\Delta_i). \tag{III.3}$$

According to (III.1) and (III.2), a parametrisation of a given sequence of laws is a generalised empirical hypothesis which, roughly speaking, implies that these laws are particular cases corresponding to suitably chosen parameter values.

2.2 Synthesis of Theories

Let $T, T_1, \ldots, T_n$ now be pairwise different, empirical theories with respective languages $L, L_1, \ldots, L_n$ and semantic models $I, I_1, \ldots, I_n$ ($n \geq 2$). Assume that the axioms of each theory comprise at least one empirical law and observe that the logical conjunction of the empirical axioms of each theory can be written in normal form.

Therefore, letting $\Lambda \in L$ and $\Omega_i \in L_i$ while keeping the form (III.1), one may assume that

$$T = C_L(\Lambda), \tag{III.4a}$$

$$T_i = C_{L_i}(\Omega_i), \quad i \leq n, \tag{III.4b}$$

without loss in generality. L' being the language of $\bigcup_{i=1}^{n} T_i \cup T$, one eventually obtains the result:

T is a unified theory with respect to $T_1, \ldots, T_n$ if the following conditions hold:

1. for every i, $i \leq n$, and descriptive constant of T_i not occurring in T, there is an L'-definition with an L-formula as its definiens so that
2. if D_i is the total of these definitions, $\bar{I}^{D_i} | T_i = I_i$, and
3. Λ parametrises $\Omega_1, \ldots, \Omega_n$ in $T' = C_{L'}(T \cup D)$, where $D = \bigcup_{i=1}^{n} D_i$.

This result, which states the synthesis of theories through parametrisation of laws, is easy to perceive. From (III.2) and

$$\wedge \xi_i \Psi [\![\eta_i, \xi_i]\!] \Rightarrow \wedge \xi_i \Phi_i [\![\xi_i]\!] \tag{III.5}$$

It follows immediately that for every i, $i \leq n$, the T_i-axiom Ω_i is an L'-consequence of the T-axiom Λ. Hence the conditions (1) and (3) imply that T includes T_i on the basis of D_i. According to (2), the expanded semantic interpretation $\bar{I}^{D_i}$ is a model of $T \cup D_i$.

The basic idea of synthesis of theories through parametrisation of laws is expressed by the simple logical fact that according to (III.5) the axioms of $T_1, \ldots, T_n$ follow from special cases of the T-axiom. This explanation scheme for intertheory relations apparently suggests itself, and numerous examples are known in the natural sciences, especially physics (Geiger 1986, 1988a,b), in which syntheses of theories or even whole disciplines have been achieved after suitable parametrisations had been found for the empirical postulates involved. However, the logical simplicity of the scheme must not mislead one about the fact that in scientific applications the synthesis of theories through parametrisation of laws may encounter considerable technical difficulties. In practice the typical situation will be that the theories to be unified are each more or less well understood or well confirmed, but their synthesis requires further research, both theoretical and experimental, in dimensions entirely unknown thus far (Geiger 1988a,b). Whether an empirically satisfying unified theory then really exists, is by no means a merely analytical problem, ignoring the question of the uniqueness of such a theory.

2.3 Informal Summary

The type of unified theory introduced above constitutes just one of many conceivable alternatives to the holism-reductionism dichotomy in philosophy and the empirical sciences. By *parameters* we understand variables occurring in (formal representations of) generalised empirical hypotheses ("laws") whose possible values vary between, rather than within, the contexts of inquiry and domains of application of these laws. A *parametrisation* of a given sequence of laws has been defined as a generalised empirical hypothesis which implies these laws as particular cases corresponding to suitably chosen parameter values. Parametrisation of the empirical axioms of two or more theories has been shown to be an instance of intertheory synthesis. These definitions and results primarily reflect the perspectives of the previous chapter, as will become apparent from Section III.4, where the terminology of system theory is resumed. But the range of applicability of the present criteria of simultaneous reduction has been argued to be much broader than system theory.

3 The Central Representation Theorem

Consider the hierarchy $\Sigma = \langle\langle M_1, S_1\rangle, \ldots, \langle M_n, S_n\rangle\rangle$ of structures with $n \geq 2$. Let $T_1, \ldots, T_n$ be again first-order theories such that for every integer r, $1 \leq r \leq n$, L_r is the language of T_r with the semantic interpretation I_r relative to $\langle M_r, S_r\rangle$. In order to construe a theory T under the semantic interpretation I reducing $T_1, \ldots, T_n$ under $I_1, \ldots, I_n$ simultaneously, a language L with the logical structure of at least an nth-order calculus would be required. In L, the definition of state and system predicates for hierarchically stratified systems would then meet with considerable syntactic difficulties. In order to render possible the application of the framework of Section III.2 to theories of hierarchical organisation, we prove our main

Representation Theorem: There are theories $D_2, \ldots, D_n$ with respective languages $L_2^D, \ldots, L_n^D$, and a function $g: \{T_2, \ldots, T_n\} \to \{D_2, \ldots, D_n\}$ with the following properties: For every r, $2 \leq r \leq n$,

1. $D_r = g(T_r)$;
2. D_r is finitely axiomatisable if T_r is, and
3. there is a semantic interpretation I_r^D of L_r^D relative to $\langle M_{r-1}, M_r \cup f(S_r)\rangle$,

$$f(S_r) = \{W \mid \vee Q(Q \in S_r \wedge W = \bigcup_{q \in Q} [\![q]\!])\} \tag{II.12}$$

so that, if I_r is a semantic model of the T_r-axioms relative to $\langle M_r, S_r\rangle$, I_r^D is a semantic model of the D_r-axioms relative to $\langle M_{r-1}, M_r \cup f(S_r)\rangle$.

The theorem states that T_r under the semantic interpretation I_r relative to $\langle M_r, S_r\rangle$ can be transformed in such a way that the resulting theory D_r again has the syntactic structure of a first-order predicate calculus, with the rank of $\langle M_r, S_r\rangle$ being reduced by one in the hierarchy. In fact, the function f generates a lower-level representation of $\langle M_r, S_r\rangle$ for $r \geq 2$, as has been noted in the context of (II.15') in Section II.3. Hence the name "Representation Theorem". Before proving the theorem, we note an immediate consequence of it (to verify by induction).

Corollary: For every r, $2 \leq r \leq n$, the transformations f and g can be iterated $r - 1$ times (symbolically, "$f^{(r-1)}$" and "$g^{(r-1)}$") with the results:

1. $\langle M_r, S_r\rangle$ is transformed into the structure $\langle M_1, \mathscr{P}(M_1) \cup S_r^* \rangle$ with $S_r^* = f^{(r-1)}(S_r)$.

2. T_r and I_r, respectively, go over into a theory D_r^* and a semantic interpretation I_r^{D*} of D_r^* relative to $\langle M_1, \mathscr{P}(M_1) \bigcup S_r^* \rangle$ so that

   a) $D_r^* = g^{(r-1)}(T_r)$;
   b) D_r^* is finitely axiomatisable if T_r is, and
   c) I_r^{D*} is a semantic model of the D_r^*-axioms relative to $\langle M_1, \mathscr{P}(M_1) \bigcup S_r^* \rangle$ if I_r satisfies the T_r-axioms relative to $\langle M_r, S_r \rangle$.

This corollary is the desired result. It uses the fact that the entire hierarchy Σ can be given a representation in terms of the one-level structure $\langle M_1, \mathscr{P}(M_1) \bigcup S^* \rangle$,

$$S^* = \bigcup_{r=1}^{n} f^{(r-1)}(S_r), \tag{III.6}$$

which is commensurable in complexity with the basic stratum $\langle M_1, S_1 \rangle$. The corollary states that the theories $T_2, \ldots, T_n$ can be transformed in such a way that the transformed theories $D_2^*, \ldots, D_n^*$ admit semantic interpretations relative to $\langle M_1, \mathscr{P}(M_1) \bigcup S^* \rangle$ under which they are true, thus preserving the material contents of $T_2, \ldots, T_n$ respectively. Morever, when a parametrisation can be found for the empirical axioms of $T_1, D_2^*, \ldots, D_n^*$, then there exists a unified theory with respect to $T_1, D_2^*, \ldots, D_n^*$. This theory not only provides a more inclusive description of Σ than $T_1, \ldots, T_n$, but is also useful in explicating those intertheory relationships with which the holism-reductionism dichotomy cannot come to grips. The subsequent sections once more deal with these issues in detail.

Tackling the proof of the Representation Theorem, we proceed in five steps. In step (A), the language L_r^D is obtained from L_r, and in step (B) we construct a function g and theories D_r, $2 \leq r \leq n$, with the property required by Part (1) of the theorem. Then Part (2) is shown in (C). In (D) and (E), we define the functions I_r^D, $2 \leq r \leq n$, and prove Part (3) of the theorem. For each step we give a proof scheme with the index variable "r". The sequence of conclusions in each scheme is independent of r, however. So we may drop "r" altogether, with $L_r, L_r^D, T_r, D_r, I_r$ and I_r^D referred to as L, L^D, T, D, I and I^D respectively. Similarly, we replace "M_{r-1}" and "S_r" by the symbols "M" and "S", assuming

$$M_r = \mathscr{P}(M_{r-1}) \tag{III.7}$$

without loss in generality. The semantic interpretations I^D and I of L^D and L are thus defined relative to the structures $\langle M, \mathscr{P}(M) \bigcup f(S) \rangle$ and $\langle \mathscr{P}(M), S \rangle$. Eventually, we resume the metatheoretic denotations and conventions of Section II.3 with obvious modifications where necessary.

(A) Given L, a suitable second-order language L_T^D is constructed, and the descriptive vocabulary of L is mapped into the descriptive vocabulary of L_T^D so that L^D can be characterised in terms of L_T^D. The syntax and logic of L, L^D and L_T^D are supposed to be specifiable in terms of any suitable first- and second-order predicate calculus respectively (e.g., Ebbinghaus et al. 1984).

If Φ is an m-place descriptive L-predicate, the symbol obtained from "Φ" by inserting the lower index "T" and the upper index "D", namely "Φ_T^D", designates an m-place second-order predicate of L_T^D ($m \geq 1$). By a similar index rule, if ξ is an individual term (constant or variable) of L, ξ_T^D is a one-place first-order predicate (constant or variable) of L_T^D. Lower-case Greek letters with upper index "D", such as "ξ^D", "η^D", etc., designate the individual L_T^D-terms.

We now introduce additional first-order L_T^D-predicate constants by suitable L_T^D-definitions, and characterise L^D as a first-order sublanguage of L_T^D. The individual constants and variables of L_T^D are the individual L^D-terms. If ξ_T^D is the one-place first-order L_T^D-predicate constant obtained from the individual L-constant ξ, ξ_T^D is also a one-place L^D-predicate. For every m-place L-predicate Φ with $m \geq 1$, we define the m-place first-order predicate constant Φ^D in L_T^D,

$$\wedge \eta_1^D \cdots \wedge \eta_m^D (\Phi^D(\eta_1^D, \ldots, \eta_m^D) \Leftrightarrow \quad (\text{III.8})$$

$$\vee \zeta_{T1}^D \cdots \vee \zeta_{Tm}^D (\Phi_T^D(\zeta_{T1}^D, \ldots, \zeta_{Tm}^D) \wedge \zeta_{T1}^D(\eta_1^D) \wedge \cdots \wedge \zeta_{Tm}^D(\eta_m^D))).$$

We denote the logical conjunction of the definitions (III.8) by "Γ", and the total of these definitions by "G", with $\Gamma \in L_T^D$ and $G \subset L_T^D$. Then we stipulate for every first-order L_T^D-predicate constant defined according to (III.8) to be an L^D-predicate, too. The meaning of (III.8) is easy to perceive. Consider the case $m = 1$, and assume "Φ_T^D" represents the set-theoretic predicate "N" of the second order. Then "Φ^D" denotes "the union of N", that is, "union of the sets contained in the class N of sets".

Finally, by the language L^D we understand the total of well-formed first-order L_T^D-formulae in which no descriptive terms occur other than the individual L^D-constants, L^D-variables and L^D-predicates. It is important to note that the above construction of the descriptive vocabulary of L^D constitutes a one-to-one correspondence between the descriptive L-constants (individual constants and primitive predicates), on the one hand, and the primitive L^D-predicates, on the other. In particular, every object constant of L corresponds to exactly one one-place predicate of L^D.

(B) For every L-sentence H, the sentence H_T^D is called the L_T^D-*substitute of* H exactly if H_T^D is obtained from H according to the following rule:

For every predicate Φ and individual term η occurring in H, all occurrences, free or bound, of Φ and η in H are replaced by occurrences of the L_T^D-predicates Φ_T^D and η_T^D respectively. The identity relation for individual terms goes over into the identity relation for one-place, first-order L_T^D-predicates.

Cum grano salis, this substitution rule preserves the syntactic form of the L-sentences. The notation "$H \leftrightsquigarrow H_T^D$" indicates that H_T^D is the L_T^D-substitute of the L-sentence H; and for every X, $X \subset L$, we define

$$X_T^D = \{ H_T^D | \vee H(H \in X \wedge H \leftrightsquigarrow H_T^D) \}. \quad (\text{III.9})$$

As customary in the literature, we use the symbol "$\vdash$" to denote provability and deducibility, "$\vdash_1$" in the metalanguages of L and L^D, and "$\vdash_2$" in the metalanguage of L_T^D. Adopting a casual, though suggestive notation, we have

$$\vdash_1 \subset \vdash_2. \quad (\text{III.10})$$

The inclusion (III.10) expresses the fact that for every pair $\langle X, \Omega \rangle$ with $X \subset L$, $\Omega \in L$, and $X \vdash_1 \Omega$, one also has $X_T^D \vdash_2 \Omega_T^D$, and for every pair $\langle X^D, \Omega^D \rangle$, $X^D \subset L^D$ and $\Omega^D \in L^D$, one has $X^D \vdash_1 \Omega^D$ if and only if $X^D \vdash_2 \Omega^D$.

Consider the function $g: \mathscr{P}(L) \to \mathscr{P}(L^D)$ which assigns the set

$$g(X) = \{ \Psi^D | \Psi^D \in L^D \wedge X_T^D \cup G \vdash_2 \Psi^D \} \quad (\text{III.11})$$

to every X, $X \subset L$. We show that $g(X)$ is a theory with the language L^D. Let $\Omega^D \in L^D$

with $g(X) \vdash_1 \Omega^D$, Ω^D being arbitrary otherwise. Because of (III.10) $g(X) \vdash_2 \Omega^D$. By virtue of (III.11) and the transitivity of $\vdash_2$, $G \bigcup X_T^D \vdash_2 \Omega^D$. Hence $\Omega^D \in g(X)$. Since the concepts of deducibility and logical consequence coincide on L^D, every L^D-consequence of $g(X)$ is an element of $g(X)$.

Since $T \in \mathcal{P}(L)$, the definition $D = g(T)$ completes the proof of Part (1) of the Representation Theorem.

(C) In order to show that D is finitely axiomatisable if T has a finite set of axioms, we use the fact that when D has a complementary theory $\hat{D}$ in L^D, it has a finite set of axioms (see, e.g., Schwabhäuser 1971, p. 155). By a theory complementary to D in L^D we mean any theory $\hat{D}$ with the language L^D for which

$$C_{L^D}(D \cup \hat{D}) = L^D \tag{III.12a}$$

$$C_{L^D}(\emptyset) = D \cap \hat{D}. \tag{III.12b}$$

Suppose T has finite set of axioms whose logical conjunction is H, and let $\hat{H}_T^D = H_T^D \wedge \Gamma$. Then

$$D = \{\Omega^D | \Omega^D \in L^D \wedge \{\hat{H}_T^D\} \vdash_2 \Omega^D\}. \tag{III.11'}$$

We show that D has the complementary theory D',

$$D' = \{\Psi^D | \Psi^D \in L^D \wedge \{\neg \hat{H}_T^D\} \vdash_2 \Psi^D\}$$

in L^D:

$$\begin{aligned}
D \bigcup D' &= \{\Omega^D | \Omega^D \in L^D \wedge ((\{\hat{H}_T^D\} \vdash_2 \Omega^D) \vee (\{\neg \hat{H}_T^D\} \vdash_2 \Omega^D))\} \\
&= \{\Omega^D | \Omega^D \in L^D \wedge ((\vdash_2 \hat{H}_T^D \Rightarrow \Omega^D) \vee (\vdash_2 \neg \hat{H}_T^D \Rightarrow \Omega^D))\} \\
&= \{\Omega^D | \Omega^D \in L^D \wedge \vdash_2 (\hat{H}_T^D \Rightarrow \Omega^D) \vee (\neg \hat{H}_T^D \Rightarrow \Omega^D)\} \\
&= \{\Omega^D | \Omega^D \in L^D \wedge \vdash_2 (\hat{H}_T^D \wedge \neg \hat{H}_T^D) \Rightarrow \Omega^D\} \\
&= L^D \\
D \bigcap D' &= \{\Omega^D | \Omega^D \in L^D \wedge (\{\hat{H}_T^D\} \vdash_2 \Omega^D) \wedge (\{\neg \hat{H}_T^D\} \vdash_2 \Omega^D)\} \\
&= \{\Omega^D | \Omega^D \in L^D \wedge (\vdash_2 \hat{H}_T^D \Rightarrow \Omega^D) \wedge (\vdash_2 \neg \hat{H}_T^D \Rightarrow \Omega^D)\} \\
&= \{\Omega^D | \Omega^D \in L^D \wedge \vdash_2 (\hat{H}_T^D \Rightarrow \Omega^D) \wedge (\neg \hat{H}_T^D \Rightarrow \Omega^D)\} \\
&= \{\Omega^D | \Omega^D \in L^D \wedge \vdash_2 (\hat{H}_T^D \vee \neg \hat{H}_T^D) \Rightarrow \Omega^D\} \\
&= \{\Omega^D | \Omega^D \in L^D \wedge \vdash_2 \Omega^D\} \\
&= \{\Omega^D | \Omega^D \in L^D \wedge \vdash_1 \Omega^D\} \\
&= C_{L^D}(\emptyset).
\end{aligned}$$

(D) The semantic interpretation I^D of L^D is a function of such a kind that

(D1) for every object constant ξ^D of L^D,

$$I^D(\xi^D) \in M;$$

(D2) if ξ_T^D is the one-place L^D-predicate obtained from the individual L-constant ξ as prescribed by (A) then

$$I^D(\xi_T^D) = I(\xi),$$

The Central Representation Theorem

where $I(\xi) \in \mathcal{P}(M)$ and, hence, $I^D(\xi_T^D) \in \mathcal{P}(M)$;

(D3) if Φ^D is the m-place L^D-predicate obtained from the m-place L-predicate Φ according to (III.8), where $m \geq 1$, $I(\Phi) = Q$ and $Q \in S$, then

$$I^D(\Phi^D) = \bigcup_{q \in Q} [\![q]\!]$$

with $I^D(\Phi^D) \in f(S)$;

(D4) I^D assigns truth values to the closed L^D-sentences in the usual manner.

The L^D-identity is semantically interpreted in I^D as customary for first-order calculi with identity.

(E) It remains to prove Part (3) of the Representation Theorem. We show this part by constructing a semantic interpretation I_T^D of L_T^D relative to $\langle\langle M, \mathcal{P}(M) \bigcup f(S)\rangle, \langle \mathcal{P}(M), S\rangle\rangle$ with the properties:

(i) If every closed T-sentence is true under I relative to $\langle \mathcal{P}(M), S\rangle$, every closed T_T^D-sentence is true under I_T^D relative to $\langle \mathcal{P}(M), S\rangle$, where T_T^D is obtained from T according to (III.9),

(ii) (III.8) is true under I_T^D relative to

$$\langle\langle M, \mathcal{P}(M)\bigcup f(S)\rangle, \langle \mathcal{P}(M), S\rangle\rangle.$$

(iii) I^D is the restriction of I_T^D to the descriptive constants and closed sentences of L^D.

From (iii) follows immediately that every closed D-sentence is true under I^D relative to $\langle M, \mathcal{P}(M)\bigcup f(S)\rangle$.

The semantic interpretation I_T^D of L_T^D is a function of such a kind that

(E1) for every object constant ξ^D of L_T^D,

$$I_T^D(\xi^D) \in M;$$

(E2) if ξ_T^D is the one-place L_T^D-predicate obtained from the individual L-constant ξ as prescribed by (A), then

$$I_T^D(\xi_T^D) = I(\xi),$$

where $I(\xi) \in \mathcal{P}(M)$;

(E3) if Φ^D is the m-place L_T^D-predicate obtained from the L-predicate Φ according to (III.8), where $m \geq 1$ and $I(\Phi) = Q$ with $Q \in S$, then

$$I_T^D(\Phi^D) = \bigcup_{q \in Q} [\![q]\!]$$

with $I_T^D(\Phi^D) \in f(S)$;

(E4) if Φ_T^D is the second-order L_T^D-predicate obtained from the L-predicate Φ according to (A), then

$$I_T^D(\Phi_T^D) = I(\Phi)$$

with $I_T^D(\Phi_T^D) \in S$;

(E5) I_T^D assigns truth values to the closed L_T^D-sentences in the usual manner.

The L_T^D-identity relations for object terms and first-order predicates are semantically interpreted in I_T^D as customary for second-order calculi.

We show that I_T^D has the property (i). For every closed T-sentence H, the L_T^D-substitute H_T^D is closed. Moreover, H and H_T^D have completely analogous syntactic forms in the sense of the substitution rule of (B), and there is a one-to-one correspondence between the descriptive constants and variables of H, on the one hand, and of H_T^D, on the other. According to (E2) and (E4), the interpretation functions I and I_T^D take on the same values for corresponding constants in H and H_T^D. Then, by the familiar Coincidence Theorem of metalogic, (i) holds.

In order to prove (ii), we must show that for every m-place L_T^D-predicate Φ_T^D of the second order, and every sequence $\langle \eta_1^D, \ldots, \eta_m^D \rangle$ of individual L_T^D-constants ($m \geq 1$), there exists a sequence $\langle \zeta_{T1}^D, \ldots, \zeta_{Tm}^D \rangle$ of one-place L_T^D-predicates of the first order with

$$\langle I_T^D(\zeta_{T1}^D), \ldots, I_T^D(\zeta_{Tm}^D) \rangle \in I_T^D(\Phi_T^D) \wedge I_T^D(\eta_1^D) \in I_T^D(\zeta_{T1}^D) \wedge \cdots \wedge I_T^D(\eta_m^D) \in I_T^D(\zeta_{Tm}^D) \tag{III.13}$$

if and only if

$$\langle I_T^D(\eta_1^D), \ldots, I_T^D(\eta_m^D) \rangle \in I_T^D(\Phi^D). \tag{III.14}$$

According to (E3) and (E4),

$$I_T^D(\Phi^D) = \bigcup_{q \in Q} [\![q]\!] \tag{III.15}$$

if $I_T^D(\Phi_T^D) = Q$. Equation (III.15) can be rewritten as

$$I_T^D(\Phi^D) = \{\langle y_1, \ldots, y_m \rangle \mid \vee q_1 \cdots \vee q_m (\langle q_1, \ldots, q_m \rangle \in Q \wedge y_1 \in q_1 \wedge \cdots \wedge y_m \in q_m)\}. \tag{III.16}$$

Let $\eta_1^D, \ldots, \eta_m^D$ now be individual L_T^D-constants, $\zeta_{T1}^D, \ldots, \zeta_{Tm}^D$ one-place L_T^D-predicate constants of the first order, and Φ_T^D a second-order L_T^D-predicate for which (III.13) holds, but which are arbitrary otherwise. Choosing

$$x_i = I_T^D(\eta_i^D) \tag{III.17a}$$

$$z_i = I_T^D(\zeta_{Ti}^D), \quad 1 \leq i \leq m, \tag{III.17b}$$

$$Q = I_T^D(\Phi_T^D) \quad \text{(as above)}, \tag{III.17c}$$

one gets from (III.13)

$$\langle z_1, \ldots, z_m \rangle \in Q \wedge x_1 \in z_1 \wedge \cdots \wedge x_m \in z_m \tag{III.18a}$$

and

$$\vee q_1 \cdots \vee q_m (\langle q_1, \ldots, q_m \rangle \in Q \wedge x_1 \in q_1 \wedge \cdots \wedge x_m \in q_m). \tag{III.18b}$$

Comparison of (III.18) with the right-hand side of (III.16) shows that

$$\langle x_1, \ldots, x_m \rangle \in I_T^D(\Phi^D). \tag{III.14'}$$

Hence (III.13) implies (III.14). In a similar fashion one demonstrates that if for given $\eta_1^D, \ldots, \eta_m^D$ and Φ_T^D there are no one-place, first-order predicate constants $\zeta_{T1}^D, \ldots, \zeta_{Tm}^D$ so that (III.18) holds, (III.14) does not hold either. This completes the proof of (ii).

In order to show that I_T^D has the property (iii), it is sufficient to compare (D1),...,(D3) with (E1),...,(E3) and (E4) respectively, and (D4) with (E5). By this remark we have completed the proof of the Representation Theorem.

3.1 The Meaning and Significance of the Theorem: An Informal Résumé

The theorem established above meets substantial theoretical challenges of the evolutionary approach to hierarchical structures and systems. On the one hand, the theorem is based on the fact that in multilevel hierarchies the upper levels of organisation admit uniquely determined lower-level representations. As a matter of principle, such representations can always be obtained, for example, by means of the transformation (II.12). An important effect of these lower-level representations is that they leave the different strata of a hierarchy commensurable in complexity. Ignorance of the commensurability problem has been argued in Section II.4 to be constitutive of major absurdities of holism.

On the other hand, the Representation Theorem shows that in a sense theories of higher-level phenomena, too, may admit lower-level representations, meaning that they may be transformed into theories of lower-level phenomena without loss in material content. In fact, this possibility has been explicitly proved for first-order theories. The strictly anti-holistic conclusion from this result is that, as long as deterministic theories are concerned, scientists may well choose microscopic or macroscopic theoretical descriptions of the real world, the difference being solely a matter of expedience and all sorts of other pragmatic criteria rather than a question of the structure of matter itself. Furthermore, the intertheory transformation characterised by the Representation Theorem renders possible both unified theories and evolutionary theories of hierarchical organisation in a straightforward fashion. Using the conceptual framework of parametrised laws, we shall exploit this possibility in the next section with regard to hierarchical systems in general, and in Part Two with regard to biosocial evolution in particular.

A few more notes regarding the significance of the Representation Theorem for evolutionary science must be added at this point. The theorem refers to theories and laws that are deterministic in the sense that they assert cause-effect implications between, rather than probabilities of, events. Restricting attention to such theories might seem inadequate to the following situation. As has been explained at length above, in hierarchically organised systems successive strata are usually termed micro- and macrolevels, respectively, of theoretical description in statistical physics, biophysics and related disciplines. Now it is well known that often the rise of spatiotemporally coherent macrostructures from microstructures inherently depends on random behaviour at the microlevel. This situation, in turn, calls for elaborate stochastic approaches and theories, which have become known as the theories of "phase transitions", "order through fluctuations" (Nicolis and Prigogine 1977; Serra et al. 1986), and "synergetics" (Haken 1978, 1983). Today these theories are widely believed to explain the fundamental motor patterns of progressive natural self-organisation in physics, biomolecular chemistry, population biology and, eventually, sociology and economics (Weidlich and Haag 1983). But, on the other hand, it is also well known that in many relevant cases deterministic theories exist and yield reasonable, approximate descriptions of evolving systems, even where stochastic effects gain critical influence (Nicolis and Prigogine 1977, Chap. 9; Babloyantz 1986). More recently developed techniques of continuum mathematics such as structural-stability approaches to deterministic dynamical systems (Thom 1975; Poston and

Stewart 1978) may then still capture important characteristics of these critical evolutionary stages. At any rate, it is to some extent a matter of the focus of interest and perspective whether deterministic or stochastic modes of description are applied to the evolutionary phenomena in point.

4 State-Determined Hierarchical Systems

In this section, we establish connexions between our notions of unified theory and parametrisation of laws, on the one hand, and the system–theoretic framework of Section II.1, on the other. In particular, we shall determine the semantic interpretations $I_1,\ldots,I_n, I$ of the theories $T_1,\ldots,T_n, T$ assumed in Section III.2 in such a way that the empirical postulates (III.1) and the sentence (III.2) are satisfied relative to the joint structure of a parameter family of input-output systems. Thereby we shall not consider the condition that T includes T_i on the basis of suitably chosen definitions since the results of the present section do not depend explicitly on this condition. In order to reduce simultaneously $T_1,\ldots,T_n$ under $I_1,\ldots,I_n$ to T under I, we rather tacitly suppose our semantic constructions to be of such a kind that this condition always holds. Special attention will be paid to the intuitive foundation of the formalism and to problems of hierarchical organisation. Again, only system theories will be considered which are elementary in the sense that $T_1,\ldots,T_n, T$ describe the action-reaction modes of real-world objects in terms of input-output systems rather than characterise the properties of such systems. For if s is a system with structure in $\langle M, S \rangle$, the properties of s are higher-order attributes not contained in S, requiring at least second-order predicates and theories for their description.

4.1 Parameter Families of Systems

Consider the structures $\langle M_1, S_1 \rangle,\ldots,\langle M_n, S_n \rangle$ and the input-output systems $s_1,\ldots,s_n$ with structures in $\langle M_1, S_1 \rangle,\ldots,\langle M_n, S_n \rangle$ respectively ($n \geq 2$). In other words, for every i, $1 \leq i \leq n$, there exists a pair $\langle X_{1i}, X_{2i} \rangle$ with $X_{1i} \in S_i$, $X_{2i} \in S_i$, and

$$s_i \subset X_{1i} \times X_{2i}. \tag{III.19}$$

Furthermore, let $\langle M, S \rangle$ be a structure such that

$$\bigcup_{i=1}^{n} M_i \subset M \quad \text{and} \quad \bigcup_{i=1}^{n} S_i \subset S. \tag{III.20}$$

The multiplicities of s_i, X_{1i} and X_{2i} are m_i, k_i and $m_i - k_i$ respectively ($m_i \geq 2, 1 \leq k_i < m_i$). Without loss in generality, suppose $s_i \in S_i$ for every i, $1 \leq i \leq n$. Although X_{1i} and X_{2i} are generally multicomponent relations and, hence, s_i can equally well be characterised in terms of the decompositions (families of components) of X_{1i} and X_{2i}, we are primarily interested in the binary form (III.19). This form not only provides a natural and simple way of describing input-output interactions and similar behaviour patterns, but also suggests itself in view of the following situation. Many empirical laws in physics and the life sciences state functional relations (i.e., functions in the mathematical sense) between observable quantities, which can be understood as input-output systems. Actually every functional relation has an input-output form irrespective of further decomposition of its input and output relations.

Still concentrating on cause-effect implications, we take the inclusion (III.19) as a basis for the conceived semantic models of the axioms (III.1) (cf. (III.4)). The assumption that these axioms describe implications equivalent to (III.19) corresponds to the following situation. Scientists are typically concerned with questions of how physical objects react upon forces induced by neighbouring objects or surrounding force fields, how organisms respond to environmental cues, and how social organisations behave under the influence of their sociocultural milieux. Now laws and theories which state generalised implications equivalent to (III.19) provide answers like this: "Any such object, organism, organisation, etc. is a system s_i with the input-output structure $\langle M_i, \{X_{1i}, X_{2i}\}\rangle$."

Let $\{R_1, \ldots, R_n\}$ be a set of relations in M, $\{R_1, \ldots, R_n\} \subset S$, with $R_i \times s_i$ having equal number m, $m > m_i$, of components for $i \leq n$, and

$$\bigcup_{i=1}^{n} (R_i \times X_{1i} \times X_{2i}) \subset S.$$

Suppose $U \in S$, where

$$U \subset \bigcup_{i=1}^{n} (R_i \times X_{1i} \times X_{2i}). \tag{III.21}$$

The inclusion (III.21) has a straightforward system–theoretic meaning. Define M_s as the total of finite sequences in M. Then for every Q, $Q \in S$, one has $Q \subset M_s$. Hence, $\langle M_s, S \rangle$ is a structure. Put $s = \bigcup_{i=1}^{n} s_i$ and $P = \bigcup_{i=1}^{n} R_i$ so that $s \subset X_1 \times X_2$, where

$$X_1 = \bigcup_{i=1}^{n} X_{1i} \quad \text{and} \quad X_2 = \bigcup_{i=1}^{n} X_{2i}.$$

Thus s is an input-output system with structure in $\langle M_s, \{X_1, X_2\}\rangle$. For every $p \in P$, define

$$\hat{s}(p) = \{z \mid z \in s \wedge \langle p, z \rangle \in U\}.$$

The family $[\hat{s}(p)]_{p \in P}$ is called a *parametrisation of s* with *parameter space P* if

$$\bigcup_{p \in P} \hat{s}(p) = s. \tag{III.22}$$

In case of (III.22), evidently

$$U = \bigcup_{p \in P} \{\{p\} \times \hat{s}(p)\} = \bigcup_{i=1}^{n} R_i \times s_i.$$

Since in this case for every $u \in U$ there is a $p \in P$ and a unique $z \in s$ so that $u = \langle p, z \rangle$, we can decompose the variable u into a state component p and a system component z in a way analogous to the decomposition of variables in Section III.2. Accordingly, P is referred to as a *global state space for s*. (In the terminology of the mathematical theory of systems of dynamic differential equations, P is also called a *control space for s*.) If and only if (III.22), we also say $[\hat{s}(p)]_{p \in P}$ *parametrises s*.

These definitions are entirely consistent with the concepts of state and state space familiar elsewhere in natural science. In fact, the behaviour and structure of physical and biological systems are often represented by input-output relations, on the one hand, and state variables ("control parameters", "structural parameters"), on the other. The distinction between system and state variables implies that a system's response (output) to an external event (input) is normally codetermined by global constraints on the system behaviour which vary independently of the inputs and outputs and are therefore lumped together under the separate concept of state. State spaces have been used to represent the environmental characteristics as well as the internal organisation of (open) systems (Mesarović and Takahara 1975, p. 255ff; Pichler 1975, Chap. 2; see also Zadeh 1969, and the references therein). But we need not distinguish between external constraints on, and internal states of, systems here since S can always be chosen so as to include parameter spaces for any kind of state or constraint of any system concerned.

4.2 Unified System Theories

We explain what we mean when we say that the empirical axiom Ω_i,

$$\Omega_i = \wedge \xi_i \Phi_i [\![\xi_i]\!], \quad i \leq n. \tag{III.1b}$$

of the theory T_i (cf. (III.4b)) asserts the inclusion (III.19). We proceed by suitably fixing a semantic interpretation I' for the language L' of $\bigcup_{i=1}^{n} T_i \cup T$ (cf. Sect. III.2) The semantic interpretations I_i and I of T_i and T, respectively, are then obtained, in a natural way, by restrictions of I' to the descriptive constants and closed sentences of T_i and T.

We suppose that the symbol "ξ_i" in (III.1b) is shorthand for the list "$\xi_{i1},\ldots,\xi_{im_i}$" of individual T_i-variables ($m_i < m$) and that for every i, $i \leq n$, there are open subsentences Φ_{i1}, Φ_{i2} and Φ_{i3} of Φ_i so that

$$\Phi_i [\![\xi_{i1},\ldots,\xi_{im_i}]\!] = (\Phi_{i1} [\![\xi_{i1},\ldots,\xi_{im_i}]\!] \tag{III.23}$$
$$\Rightarrow \Phi_{i2} [\![\xi_{i1},\ldots,\xi_{ik_i}]\!] \wedge \Phi_{i3} [\![\xi_{i,k_i+1},\ldots,\xi_{im_i}]\!]).$$

According to the denotational conventions of Section II.3, the numbers of free variables in Φ_{i1}, Φ_{i2} and Φ_{i3} are m_i, k_i and $m_i - k_i$ respectively (meaning that no variables occur free in the subsentences of (III.23) other than those that are made explicit). For every sequence $\langle \theta_{i1},\ldots,\theta_{im_i} \rangle$ of individual T_i-constants, we determine

$$I'(\Phi_{i1} [\![\theta_{i1},\ldots,\theta_{im_i}]\!]) = true$$
$$\Leftrightarrow \langle I'(\theta_{i1}),\ldots,I'(\theta_{im_i}) \rangle \in S_i,$$
$$I'(\Phi_{i2} [\![\theta_{i1},\ldots,\theta_{ik_i}]\!]) = true$$
$$\Leftrightarrow \langle I'(\theta_{i1}),\ldots,I'(\theta_{ik_i}) \rangle \in X_{1i},$$
$$I'(\Phi_{i3} [\![\theta_{i,k_i+1},\ldots,\theta_{im_i}]\!]) = true$$
$$\Leftrightarrow \langle I'(\theta_{i,k_i+1}),\ldots,I'(\theta_{im_i}) \rangle \in X_{2i}.$$

Because of (III.19) and (III.23) I' is indeed a semantic model of the empirical axioms $\Omega_1,\ldots,\Omega_n$.

In a similar fashion, I' can be further specified so as to satisfy Λ,

$$\Lambda = \wedge \zeta \Psi[\![\zeta]\!] \tag{III.1a}$$

Let "ζ" in (III.1a) be shorthand for the list "$\zeta_1,\ldots,\zeta_m$" of individual T-variables, and assume open subsentences Ψ_1 and Ψ_2 of Ψ so that

$$\Psi[\![\zeta_1,\ldots,\zeta_m]\!] = (\Psi_1[\![\zeta_1,\ldots,\zeta_m]\!] \Rightarrow \Psi_2[\![\zeta_1,\ldots,\zeta_m]\!]). \tag{III.24}$$

For every sequence $\langle \theta_1,\ldots,\theta_m \rangle$ of individual T-constants, choose

$$I'(\Psi_1[\![\theta_1,\ldots,\theta_m]\!]) = true$$
$$\Leftrightarrow \langle I'(\theta_1),\ldots,I'(\theta_m) \rangle \in U$$

$$I'(\Psi_2[\![\theta_1,\ldots,\theta_m]\!]) = true$$
$$\Leftrightarrow \langle I'(\theta_1),\ldots,I'(\theta_m) \rangle \in \bigcup_{i=1}^{n} (R_i \times X_{1i} \times X_{2i}).$$

Because of (III.21) and (III.24) I' satisfies Λ relative to $\langle M, S \rangle$.

We finally specify I' so as to satisfy the condition

$$\wedge \xi_i (\Psi[\![\eta_i, \xi_i]\!] \Rightarrow \Phi_i[\![\xi_i]\!]) \tag{III.2}$$

relative to $\langle M, S \rangle$. In this semantic interpretation, the unified theory T is then the theory of a parameter family of systems with structures in $\langle M, S \rangle$. Let $s_1,\ldots,s_n$ and s be given as above, and let $[\hat{s}(p)]_{p \in P}$ parametrise s. Assume a subset $\{p_1,\ldots,p_n\} \subset P$ so that

$$p_i \in R_i \tag{III.25a}$$

$$p_i = \langle I'(\eta_{i1}),\ldots,I'(\eta_{i,m-m_i}) \rangle \tag{III.25b}$$

$$s_i = \hat{s}(p_i), \quad 1 \leq i \leq n, \tag{III.25c}$$

where the symbol "η_i" occurring in (III.2) is shorthand for the list "$\eta_{i1},\ldots,\eta_{i,m-m_i}$" of individual T-constants. Then (III.2) is clearly satisfied under I' relative to $\langle M, S \rangle$. Observe that because of (III.25c) the generalised implication (III.2) remains true under I' even if "$\Rightarrow$" is replaced by "$\Leftrightarrow$" in it.

The preceding semantic constructions, though perhaps lengthy, are by no means contrived. Their meaning is indeed straightforward to rephrase intuitively. Given any systems $s_1,\ldots,s_n$, the questions arise whether and how these systems may be interrelated. If (III.25) holds, the following answer can be given: There is a system s with state space P so that s has exactly the properties of s_i in the state p_i ($i \leq n$). In other words, the systems $s_1,\ldots,s_n$ are state-determined subrelations of s. This explanation scheme of intersystemic relationships virtually guides the argument of the entire book. In particular, Part Two is concerned with applications to social, biological and cultural systems, and the evolutionary relationships between them.

4.3 State-Determined Hierarchies

Parameter representations can be applied to characterise stratified systems, too. This possibility is particularly instructive with regard to what has been called the hierarchical organisation of matter. Suppose once more that s is an input-output system with parameter space P and parametrisation $[\hat{s}(p)]_{p\in P}$. Let $q\in P$ so that $\hat{s}(q)$ is a coupled system of coupling degree $d_c \geqq 2$ in $[\hat{s}(p)]_{p\in P}$. Then, according to Section II.1, there exists a $(d_c + 1)$-level hierarchy C of coupling relations based on $[\hat{s}(p)]_{p\in P}$. We refer to C as a *state-determined stratified system* because the coupled and decoupled subsystems of $\hat{s}(q)$ are state-determined subrelations of s, and their coupling degrees correspond uniquely to the strata of C.

The concept of a state-determined stratified system has been illustrated elsewhere (Geiger 1986) by an example from physics (plasma state, monoatomic and molecular gas, fluid and solid state hydrogen). The example involves two four-level hierarchies, each associated with four aggregation states of the most elementary chemical substance; the relevant state variable is temperature. A less sketchy treatment of a more modest example will be given in the Appendix.

State-determined hierarchies of systems offer useful epistemic perspectives on natural self-organisation. Above all, they refute the naive holistic contention that the hierarchical structure of matter prescribes us the scientific theories, and these theories' semantic interpretations, in unique, doctrine-like manners. Rather, the hierarchical organisation of many real-world systems depends on empirically contingent coupling states of one-level systems. Whether one then prefers to interpret one's theories relative to multilevel or one-level structures may largely become a matter of the availability of suitable parameter representations for the systems and hierarchies concerned. As a matter of principle, however, one-level representations of hierarchical structures are always possible, as the Representation Theorem shows. Thus approaches to natural self-organisation do exist which suggest that hierarchical complexity does not pose epistemic problems other than those familiar from more narrowly circumscribed physical and organic phenomena.

4.4 Evolutionary Processes

Parameter representations of systems and empirical laws are useful in characterising evolutionary processes and theories. In particular, they may render possible complete, deterministic descriptions of processes involving increases in hierarchical complexity. So any such description can be assigned to exactly those properties that the doctrine of emergent evolution denies an evolutionary theory can have.

Let $\mathcal{K}$ be a completely ordered subset of the parameter space P, with some order relation $\underline{\leq}$. The horizontal bars in "$\underline{\leq}$" indicate that the identity is a subrelation of $\leq$, while for any two elements $p\in\mathcal{K}$, $p'\in\mathcal{K}$ either $p \prec p'$ or $p' \prec p$ if $p \neq p'$. In many applications of the present formalism, P and $\mathcal{K}$ will be (topologically isomorphic to) sets of real numbers or real intervals. For instance, quantitative concepts of time compatible with $\underline{\leq}$ may be defined with reference to $\mathcal{K}$, or $\mathcal{K}$ may be some external constraint (e.g., energy inflow or temperature) imposed on a physical or organic system. In the former case, $\mathcal{K}$ is referred to as a *time set*. For every pair $\langle p', p''\rangle \in \mathcal{K}^2$ with $p' \underline{\leq} p''$, the set

$$\mathcal{K}_{p'p''} = \{p | p\in\mathcal{K} \wedge p' \underline{\leq} p \underline{\leq} p''\}$$

is called a *segment* of $\mathcal{K}$ (e.g., a time segment). For every pair $\langle p', p''\rangle$ with $p' \underline{\leq} p''$ and segment $\mathcal{K}_{p'p''} \subset \mathcal{K}$, the set

$$\pi(\mathcal{K}_{p'p''}) = \{\langle p, \hat{s}(p)\rangle | p\in\mathcal{K}_{p'p''}\}$$

State-Determined Hierarchical Systems

is referred to as a *process on* $\mathcal{K}_{p'p''}$. Briefly and informally, a process is a sequence of states a system may take on one after another under changing constraints. The process $\pi(\mathcal{K}_{p'p''})$ on the segment $\mathcal{K}_{p'p''}$ is called an *evolutionary process* if and only if $\hat{s}(p'')$ is behaviourally or structurally more complex than $\hat{s}(p')$. If so, we also say that $\hat{s}(p'')$ *evolves from* $\hat{s}(p')$ *in the course of* $\pi(\mathcal{K}_{p'p''})$. In cases in which $\hat{s}(p'')$ is indeed behaviourally more, but structurally less, complex than $\hat{s}(p')$, and in similar situations, we have to be more specific about the nature of evolutionary processes. For instance, we may then say that $\pi(\mathcal{K}_{p'p''})$ is an *evolutionary process with respect to behavioural complexity*, and so forth.

Let d'_c and d''_c be the coupling degrees of $\hat{s}(p')$ and $\hat{s}(p'')$ in $[\hat{s}(p)]_{p\in P}$. Then $\pi(\mathcal{K}_{p'p''})$ is an *evolutionary process with respect to hierarchical complexity* if $d'_c < d''_c$. If so, then $\pi(\mathcal{K}_{p'p''})$ is also an evolutionary process with respect to structural complexity by definition of the concepts of structural complexity and coupling degree of coupled systems. The present concept of an evolutionary process with respect to hierarchical complexity means that there is a state-determined hierarchical system based on $[\hat{s}(p)]_{p\in P}$ whose number of strata is increased from $d'_c + 1$ to $d''_c + 1$ in the course of $\pi(\mathcal{K}_{p'p''})$. The point whether such an increase in hierarchical complexity occurs or not depends exclusively on the process $\pi(\mathcal{K}_{p'p''})$. Emergent attributes, that is, subrelations of s that are not determined by $\pi(\mathcal{K}_{p'p''})$ do not arise when s evolves from p' into p''.

In fact, the following criterion is clearly necessary and sufficient for the occurrence of two or more hierarchically ordered modes of interaction among the systems in $[\hat{s}(p)]_{p\in P}$: There is a subset $V = \{p', p'', q_1, \ldots, q_k\}$ of P so that for $k \geq 2$

1. $\hat{s}(p')$ is a system component of the trivial coupling $\overset{k}{\underset{l=1}{\boxtimes}} \hat{s}(q_l)$, whereas
2. $\hat{s}(p'')$ is a non-trivial coupling of $\hat{s}(q_1), \ldots, \hat{s}(q_k)$.

From (1) follows

$$d'_c = 0 \qquad (III.26a)$$

for the coupling degree d'_c of $\hat{s}(p')$ in $[\hat{s}(p)]_{p\in V}$; but (1) also implies that the subsystems $\hat{s}(q_1), \ldots, \hat{s}(q_k)$ do not interact when s is in the state p'. Condition (2) entails

$$d''_c \geq 1 \qquad (III.26b)$$

for the coupling degree d''_c of $\hat{s}(p'')$ in $[\hat{s}(p)]_{p\in V}$. The criterion altogether means that in the state p'' interactions occur between the subsystems of $\hat{s}(p'')$ that do not arise in the state p'. In other words, the state transition from p' to p'' consists in turning non-interacting systems in $[\hat{s}(p)]_{p\in P}$ into interacting ones. This correspondence between states and couplings of systems is clearly incompatible with the notion of an emergent attribute. Once a parametrisation has been found for a given input-output system s, the associate parameter family not only gives the state-determined subrelations of s. It also determines whether, and which, relations (interactions, couplings, etc.) exist between any two or more systems contained in it. The state representation of s thus admits a complete, deterministic account of what happens to previously decoupled systems when they enter into interaction. Although interconnexion "confers novel properties on systems" (to use the holistic phrase), the disclosure of such properties is no more than a deterministic effect exclusively due to the attributes, and their dependence on the state, of the systems involved. This correspondence between states and interconnexions of systems is illustrated by a simple example of ecological interaction in the Appendix.

4.5 Informal Summary and Discussion

The conceptual framework of parametrised laws and theories synthesis has been employed to construct, in model-theoretic terms, a unified theory of a parameter family of input-output systems. The construction is indeed very simple, but exemplifies important corollaries to the anti-holistic conclusions of the present chapter. It has been demonstrated that systems hierarchies and evolutionary processes involving increases in hierarchical complexity may need no more than first-order deterministic theories for their description. This result depends on two conditions. Firstly, in hierarchical systems and structures, the higher levels of organisation, and their theories, can be given lower-level representations without loss in content. Secondly, suitable parameter representations can be found for the systems and hierarchies concerned. By the Representation Theorem of Section III.3, the first condition indeed holds for first-order theories and their semantic structures. The second condition is exclusively a matter of system-theoretic conceptualisation and empirical research, having countless positive instances in virtually all empirical sciences using system concepts. Thus approaches to natural self-organisation do exist which suggest that hierarchical complexity and evolution pose no other epistemic problems than are already known from more narrowly circumscribed physical and organic structures and processes.

A few comments on the relationship between the above results and the examples studied below should be made here. The parameter families and hierarchies of systems treated in the Appendix and in Part Two consist of continuous-time dynamical systems. Now the concept of continuity is a concept of mathematical topology theory which, on its part, is well known to be a higher-order theory. So the examples of a parameter family and evolutionary process introduced in the Appendix require more complicated theories than those considered above. However, this discrepancy is irrelevant to the argument of the present section. It is easy to switch from the continuous-time dynamical systems discussed in the Appendix to the associate discrete-time systems which to some considerable extent admit formalisations in terms of first-order theories. In fact, the parametrised, discrete-time dynamical systems of population genetics are given a first-order theoretical description in Geiger (1988b).

5 Pointless Scientific Controversies

Often it is the parametrisation of laws by which previously uncertain and controversial intertheory relationships can eventually be cleared up. As a first example, consider the frequent case of two theories which describe similar phenomena or share major parts of their empirical contents. Put in technical terms, the available observational facts indicate that the structures relative to which the theories are semantically interpreted have finite isomorphic substructures, or have finite substructures in common. The empirical data are thus compatible with the conjecture that the two theories have isomorphic semantic models, or that one of the models is a submodel of the other. These are typical situations in which some of the experts suspect one of the theories to be reducible to the other, or one of the structures to be an elementary substructure of the other. Yet other scientists will object that phenomenological similarities between real-world structures often arise incidentally and, therefore, bear no theoretical significance per se.

In controversies of this kind, the opponents tend to ignore the possibility of alternative intertheory relationships which may render the debates altogether pointless. In order to outline such an alternative, consider two relations s' and s'' in the universes of the two theories, respectively, and suppose s' and s'' to be isomorphic or to have subrelations is common ($s' \cap s'' \neq \emptyset$). Assume another theory T, and relations P and s falling within the range of the semantic interpretation of T. As an empirical postulate, the theory T specificies a function $\hat{s}: P \to \mathcal{P}(s)$ of such a kind that the family $[\hat{s}(p)]_{p \in P}$ parametrises s, and s' and s'' are elements of $[\hat{s}(p)]_{p \in P}$. Then the problem of which theoretical significance the observed similarities bear, can be resolved. What s' and s'' have, and what they do not have, in common is exclusively determined by P, s and the function $\hat{s}$ according to T. So the phenomenological similarities in question need not be incidental; they may well turn out to be deterministic effects in some sufficiently inclusive theoretical description. On the other hand, neither of the initial theories implies or reduces the other theory because neither contains the law that explicates the empirical relationship between s, s' and s''. Intertheory relations of this kind typically arise in theoretical population biology, when different systems of interacting populations give arise, in parameter-dependent ways, to similar forms of dynamic behaviour (e.g., asymptotic stability and homeostatic oscillations).

5.1 Cross-Level Similarity of Structures

There is an important issue in the holism-reductionism dabate in evolutionary biology, especially sociobiology, which belongs in this context. It is the theoretical significance of phenomenological similarities between different levels of organisation in biological structures and systems. (Sometimes cross-level similarity relationships are also called "interlevel analogies"; we shall review various explications and applications of the notion of similarity briefly below.) Referring to their concepts of emergence and emergent evolution, holists deny any theoretical significance of the interlevel similarities whatsoever.

We outline two possible interpretations of the holistic standpoint. The first is based on the principle (A) of holism in the strong sense (see Sect. II.3.4). According to this version of holism, interlevel relationships, including isomorphisms or any other types of analogy or resemblance, are never causal effects. As to sociobiology, for example, holists believe individual adaptations to the social structure to be attributes totally different from group adaptations because interactions between individuals, and interactions between groups (social systems) of individuals are hierarchically ordered. For the biological implications of these issues see Williams (1974), Wilson (1980), Dawkins (1982, Chaps. 5, 6) and Gould (1982). The holists insist on the point that adaptive evolution at the individual and at the group level has different causes and different, though similar, effects and, therefore, constitutes alternative biological processes. Hence the holistic notion that interlevel similarities are fortuitous effects, contributing nothing to the material contents of theories of hierarchical organisation.

In Section II.3, we have shown that this sort of holistic reasoning is generally incorrect. Irrespective of their resemblance or disparity, the phenomena characteristic of different levels of organisation in a hierarchy may well be linked in regular, deterministic ways. If so, they constitute cross-level empirical relations which are the essence and focus of interest of any attempt to account for the hierarchy in evolutionary terms. In particular, when a hierarchically organised system admits a state parametrisation, the evolution of interlevel similarities – just as any other cross-level relations, too – can be completely explained by variations in state. After all, population biology provides many examples in which state parameters exist for the selection mechanisms in animal groups, with individual adaptations causing group adaptations (i.e., increasing population numbers) in state-determined ways (Ginzburg 1977; Roughgarden 1979, Chap. 17; Wilson 1980, pp. 96–97).

5.2 Cultural Systems as "Emergent Wholes"

We now consider the problem of the cross-level phenomenological parallels with respect to another version of the holistic principle according to which coupled systems ("wholes") have emergent attributes as compared to the attributes of their decoupled subsystems ("parts"). In Section II.3, this version has been correctly rephrased as "coupled systems are defined in terms of structures that are emergent relative to the defining structures of their decoupled subsystems". Holism in the sense of this principle has fuelled much controversy on the applicability of biological laws and theories to human social patterns (Barkow 1980; Corning 1983, pp. 80–84; Rose et al. 1984, Chap. 10). On the one hand, many ethologists and sociobiologists conjecture the behavioural and social structures, which are similar in animals and man, to arise from adaptive, heritable behavioural dispositions. These biologists insist on the point that almost all heritable organic traits, whether in animals or man, are subject to natural selection, according to neo-Darwinian theory. Prominent examples of cross-species behavioural similarities are those patterns of social cooperation and competition which the sociobiologists maintain to vary within and between populations in adaptive ways, be it under ecological stress or in sociocultural contexts (e.g., of the economic market). Correspondingly, the biological explanations of human sociocultural variation have been subsumed under the hypothesis that culture tends to simulate biological adaptations (Barash 1977, Chap. 10; Alexander 1979, Chap. 2; Irons 1979). On the other hand, there are numerous holists among the opponents of human sociobiology who raise the following argument: Cultural systems are coupled systems, or "wholes", arising from the recurrent interactions of large numbers of human individuals viewed as coupled subsystems, or "parts". Similarities between animal and human behaviour bear no theoretical relevance. For the evolution of animal behaviour is an effect of natural selection which acts on the individual organism, whereas human behavioural traits are conferred upon individuals by the sociocultural system "as a whole" (learning, tradition, socialisation).

In order to assess the opposing views in this variant of the holism-reductionism controversy, we briefly consider three different connotations of the term "similarity". Then we indicate the significance of one of these connotations, namely system simulation, with regard to applications in science and engineering. On the basis of the simulation concept, we finally outline an alternative to the opposing views in the controversy.

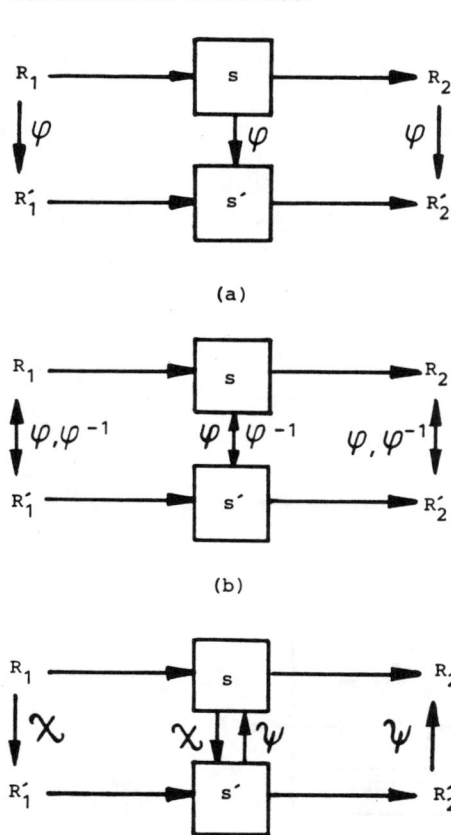

Fig. III.2a–c. Types of similarity between the input-output systems s and s'. **a** Homomorphic similarity with homomorphism φ; **b** isomorphic similarity with isomorphism φ and inverse φ^{-1}; **c** simulation of s and s' with simulation mappings $\chi: R_1 \to R'_1$ and $\psi: R'_2 \to R_2$

5.3 System Simulation

Let s and s' be systems with the structures $\langle M, \{R_1, R_2\}\rangle$ and $\langle M', \{R'_1, R'_2\}\rangle$, respectively, and assume a function $\varphi: R_1 \cup R_2 \to R'_1 \cup R'_2$. Figure III.2 illustrates three system-theoretic explications of the notion of similarity of s to s'. If for every x and y with $\langle x, y\rangle \in s$ one has $\langle \varphi(x), \varphi(y)\rangle \in s'$, then s and s' are called *homomorphically similar*, and *isomorphically similar* if φ maps s onto s', and the inverse function φ^{-1} exists and throws s' onto s. Let $\chi: R_1 \to R'_1$ and $\Psi: R'_2 \to R_2$. Then s' is said to *simulate* s if and only if

$$\wedge x \wedge y (\langle x, y\rangle \in s \Rightarrow \vee z (\langle \chi(x), z\rangle \in s' \wedge \Psi(z) = y)).$$

(s' is a *simulation system of* s).

Simulation systems have wide fields of application in science and engineering (Pichler 1975). There are several aspects of system simulation that are relevant to our analysis of the holistic principle and its application to evolutionary processes. Firstly, the coupling of systems is a frequent and effective technique to construct simulation

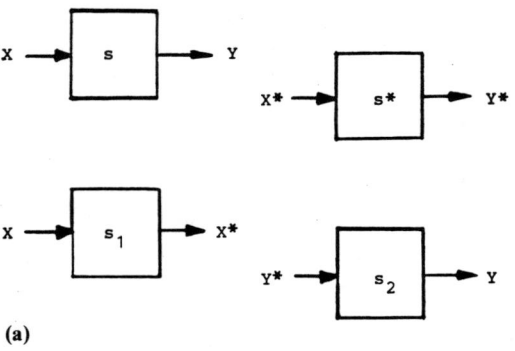

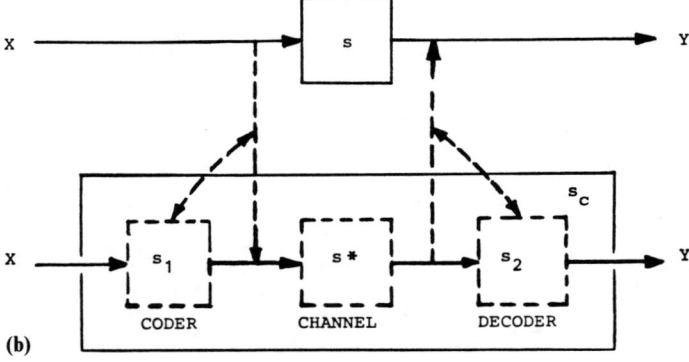

Fig. III.3a, b. Simulation of decoupled systems through coupled subsystems: example of a communication channel. **a** Decoupled systems s, s^* (channels), s_1 (coder) and s_2 (decoder). **b** Communication system s_c with coupled subsystems s^*, s_1 and s_2. If s_1, s_2 are functional systems and $s = s_c$, then s_1 and s_2 act as simulation functions (cf. Fig. III.2c), and s^* simulates s. A typical application of the schematic arrangement (**b**) is the modulation of messages by means of modulating and demodulating devices (s_1 and s_2 respectively) so as to form from messages signals having characteristics suited to the information channel s^* (e.g., Rosie 1973, Chap. 8)

systems for all sorts of purposes in system science and engineering. To confer a desired property upon some given system s, one may connect s with suitably chosen systems s_1 and s_2 to get a composite system s_c of the following kind: (i) s is transformed into the coupled subsystem s^* of s_c; (ii) s^* simulates s; (iii) s^* has the desired property (Fig. III.3). Secondly, the coupling of systems for simulation purposes is often accomplished on the basis of parameter representations of the systems involved. Coupled subsystems then correlate, in state-determined manners, with those decoupled subsystems that they simulate. Correlations of this type are characterised by (III.26). Figure III.3 illustrates the correspondence between the coupling and simulation of systems in terms of an example from communication engineering. Thirdly, the possibility of simulating systems by connecting them with other systems refutes the holistic dogma that previously non-interacting ("isolated") systems acquire emergent attributes when entering into interaction. Here "emergence" means that resemblance between the decoupled and coupled subsystems of a given system is a fortuitous, as compared to causal, effect. Contrary to the holistic dogma, compound systems do exist so that, as a state-determined effect, their coupled subsystems simulate (are similar to) those previously "isolated parts" from which the "whole" is formed.

5.4 The Problem of Similarity in Comparative Ethology

These results can also be applied to the controversial question of the meaning of similarity in comparative studies of animal and human behaviour. The problem is whether enculturation and socialisation of human individuals act so as to adapt humans to their sociocultural environments in ways predicted by neo-Darwinian adaptation theory ("culture simulates biological adaptations"). In our analysis, we concentrate on those variations in human social behaviour to which neither the adaptive hypotheses of sociobiology nor the opposing holistic views seem to apply. A case in point is ritualisation in animal and human behaviour. Although ethologists have shown that ritual behaviour in man shares basic, universal features with that in non-human animals (Eibl-Eibesfeldt 1979, 1984, Chap. 6.4), there are also human ritual expressions that vary cross-culturally in apparently non-adaptive ways. Now these variations suggest the hypothesis that humans tend to utilise biological response patterns to establish culture-specific social relations. This hypothesis is supported by the observation that human social actors sometimes encode their culturally, historically contingent wishes, needs and goals in biological terms. As for the example of human ritual, ritualised social practices apparently become easier to exploit for sociohistorically contingent purposes (e.g., commercialised in modern societies) the more firmly they are rooted in the biological, phylogenetic heritage of man (cf. Eibl-Eibesfeldt 1984, p. 639). Similarity between cultural and organic structures may thus prove an effect of human systems engineering. In this sense, simulating sociocultural through biosocial relations is a technique to confer on human individuals and groups culturally preferred attributes that are characteristic of biological systems. Explanations of sociocultural structures in biological terms which follow this line of reasoning are elaborated upon in Chapter VI.

5.5 Darwinism Versus Mendelism

As a final example of controversial intertheory relations, consider two or more theories $T_1,\ldots,T_n$ with the language L of $\bigcup_{i=1}^{n} T_i$ and the semantic interpretation I of L relative to some given domain of real objects. Under the interpretation I, each of the theories is, to some extent, supported by observations. But it is also found that $\bigcup_{i=1}^{n} T_i$ is logically inconsistent. Then the controversy concerns the question as to which of the theories is solely acceptable. This was indeed the situation in evolutionary biology after the rediscovery of Mendel's laws of genetic inheritance in 1900. With Mendelian transmission of organic traits, polymorphism is *ceteris paribus* maintained in plant and animal populations over generations, whereas it tends to decrease under

Darwinian natural selection. Mendel's and Darwin's theories were both believed by their proponents to apply to the same phenomena, and for either theory there was limited empirical evidence. But the observational data in support of Mendel's laws were difficult to reconcile with Darwin's theory, and conversely. These and other apparent inconsistencies between the proposed laws of organismic reproduction played an important role in the bitter controversy between "Mendelians" and "Darwinians" during the first decade of this century (Ewens 1979, Chap. 1; Bowler 1984, pp . 256–265). The dilemma was eventually resolved by the establishment of a neo-Darwinian synthetic theory which combines the principles of genetic inheritance with those of natural selection.

Scientific controversies of the kind *Mendelism versus Darwinism* tend to arise from theoretical generalisations that are based on observational evidence too limited to justify general hypotheses. Inconsistencies which may result in this way can sometimes be removed by suitably weakening the axioms of $T_1,\ldots,T_n$. This possibility has been exemplified for $n=2$ elsewhere (Geiger 1988b). The empirical postulates of T_1 (Mendel) and T_2 (Darwin) can be constrained so as to admit a parametrisation in T (neo-Darwinian theory). The constrained postulates also give rise to subtheories T'_1 and T'_2 of T with the same languages as T. So every semantic model of T is a model of T'_1 and T'_2, too, and T synthesises T'_1 and T'_2.

Appendix: Examples

1 Interacting Biological Populations

We introduce the concept of dynamical system to illustrate the notions of parametrisation of systems, evolutionary process, and state-determined hierarchy.

1.1 Dynamical Systems

Roughly speaking, a dynamical system is a system in which the input-output behaviour changes in time in such a way that, given the input at some time t, the system response at any time later than t is uniquely determined by the state of the system at that later time. If $\langle M, S \rangle$ is a structure and M endowed with a topology, the dynamics of a system with structure in $\langle M, S \rangle$ may be classified as being discrete or continuous in time with reference to this topology. Since the dynamical systems considered in Part Two are all systems of real differential equations, we restrict our explicit definition of the term "dynamical system" to continuous-time systems with real-valued input and output variables. More abstract and comprehensive treatments of dynamical systems can be found in Pichler (1975), and Mesarović and Takahara (1975).

Let X be a non-empty subset of $\mathbb{R}^m$ ($m \geq 1$), $\sigma: \mathbb{R} \to \mathscr{P}(X^2)$ a function, and "t" a real parameter which, in applications in science and engineering, denotes time. Then the relation s,

$$s = \bigcup_{t \in \mathbb{R}} \sigma(t) \tag{A.1}$$

with $s \subset X^2$ is called a *dynamical system* (*associated with* the parameter family $[\sigma(t)]_{t \in \mathbb{R}}$) if and only if

1. for every t, $t \in \mathbb{R}$, the relation $\sigma(t) \subset X^2$ is functional, i.e., there is a function $\sigma_t: X \to X$ so that
$$\wedge x \wedge y(y = \sigma_t(x) \Leftrightarrow \langle x, y \rangle \in \sigma(t));$$
2. σ_0 is the identical mapping on X, i.e., $\sigma_0(x) = x$ for every $x \in X$;
3. $\sigma_{t''}(\sigma_{t'}(x)) = \sigma_{t'+t''}(x)$ for every $x \in X$, $t' \in \mathbb{R}$ and $t'' \in \mathbb{R}$, and
4. the function assigning $\sigma_t(x)$ to the pair $\langle t, x \rangle \in \mathbb{R} \times X$ is smooth (i.e., continuously differentiable).

Condition (1) implies that the input-output behaviour of a dynamical system, though varying with time, remains functional in every state of the system (i.e., at any time t).

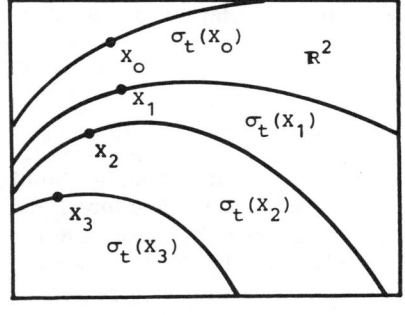
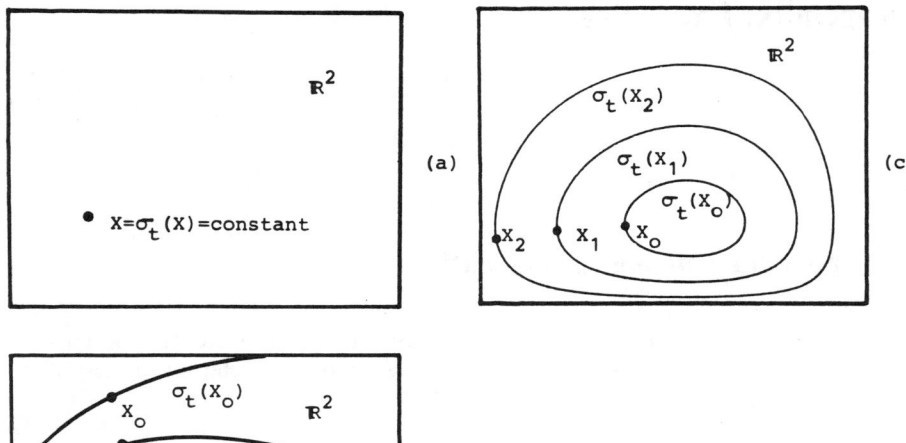

Fig. A.1a–c. Schematic representation of dynamic trajectories in the phase space $\mathbb{R}^2$. **a** Equilibrium; **b** phase portrait with non-intersecting curves, which may converge or diverge, however; **c** phase portrait with periodic orbits

The meaning of (3) can be circumscribed as follows. Assume that at time $t = 0$ the system is fed with the input x to which it responds with the output y at time t', $t' > 0$. Then the system response at some later time $t' + t''$, $t'' > 0$, is the same as the response to the input y at $t = 0$ would be at time t''. More loosely, the property (3) implies that the time dependence of the input-output behaviour of s is the same for the intervals $[0, t'']$ and $[t', t' + t'']$. Condition (2) simply specifies this time dependence for the trivial case of an interval of length zero. The final condition (4) constitutes what is referred to as continous-time dynamics.

It is well known (cf. Hirsch and Smale 1974, Chap. 8) that for every smooth function $F: X \to X$ the system

$$\frac{dZ(t)}{dt} = F(Z(t)) \tag{A.2}$$

of ordinary time-dependent differential equations corresponds uniquely to a dynamical system $s \subset \mathbb{R}^m \times \mathbb{R}^m$ associated with the parameter family $[\sigma(t)]_{t \in \mathbb{R}}$ according to

$$\wedge t \wedge x \wedge y (\langle x, y \rangle \in \sigma(t) \Leftrightarrow x = Z(0) \wedge y = Z(t)). \tag{A.3}$$

Note that the function F in (A.2) does not depend explicitly on t ("autonomous" dynamical system), as is the case in all our applications. The set X is customarily called the *phase space* of the dynamical system s, its elements being *phase points*, with the *dynamic variable* $Z(t)$ ranging over the phase space. A phase point y with $F(y) = 0$ is an *equilibrium*, or *stationary point*, of s. For every x, $x \in X$, $\sigma_t(x)$ is called a *trajectory* of the system when considered as a function of time t, while the family $[\sigma_t(x)]_{x \in X}$ of

Interacting Biological Populations

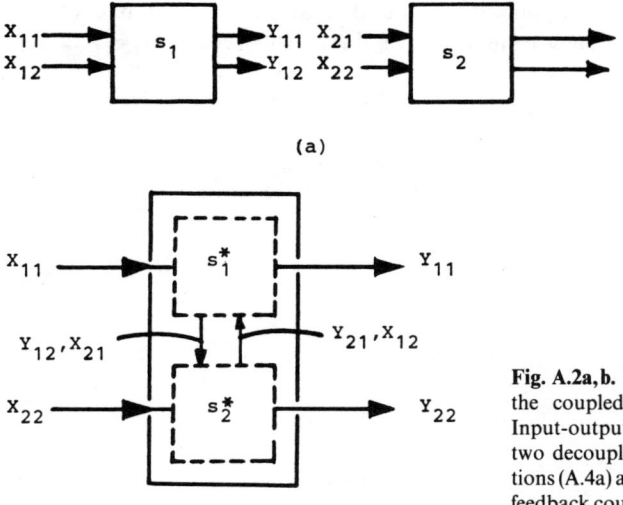

Fig. A.2a,b. Schematic representation of the coupled dynamical system (A.4). **a** Input-output systems corresponding to the two decoupled dynamic differential equations (A.4a) and (A.4b) with $\beta = 0$. **b** Double feedback coupling of s_1 and s_2 with coupled subsystems s_1^* and s_2^* for $\beta \neq 0$

trajectories is referred to as the *phase portrait* of s. The three possible types of trajectories which a dynamical system can have are schematically shown in Fig. A.1. Observe that non-periodic intersecting trajectories do not occur. This situation corresponds to an important uniqueness property of (autonomous) dynamical systems, namely that for every $x \in X$, there exists exactly one trajectory passing through x (cf. Hirsch and Smale 1974, Chap. 8).

1.2 Interacting Populations as Coupled Systems

As an example of (A.2), consider the system

$$\frac{dZ_1(t)}{dt} = k_1 Z_1(t)(K_1 - Z_1(t) - \beta Z_2(t)) \tag{A.4a}$$

$$\frac{dZ_2(t)}{dt} = k_2 Z_2(t)(K_2 - Z_2(t) - \beta Z_1(t)) \tag{A.4b}$$

which contains the familiar population-dynamic equations of two species in ecological competition. Its biological background will be explained in more detail in Section V.2. The parameter "β" designates the degree of ecological niche overlap ($0 \leq \beta \leq 1$), while for $i = 1, 2$ the function Z_i specifies the time-dependent population number ($0 \leq Z_i(t)$ for every time t), the parameter "K_i" designates the equilibrium population number for non-interacting species ($\beta = 0$, $K_i \geq 0$), and k_i is the rate constant of population growth ($k_i \geq 0$).

Figure A.2 gives the system (A.4) a graphical representation, whereby competitive interaction between the two species is treated as a coupling relation between population-dynamical systems. Either species is conceived of as such a system with two input and two output components (Fig. A.2a). For $i = 1, 2$ there is one input component, X_{ii}, for the initial population number at time t, and one output component, Y_{ii}, for the dynamic response $Z_i(t')$, $t' \geq t$. Apparently, $X_{ii} = Y_{ii} \subset \mathbb{R}_0^+$, where

$$\mathbb{R}_0^+ = \{x | x \in \mathbb{R} \wedge 0 \leq x\}.$$

The other input-output components X_{ij}, Y_{ji} arise from the coupling terms proportional to β in (A.4), where $i = 1, 2$ and $j = 1, 2$ with $i \neq j$. Accordingly,

$$X_{ij} = \{\beta Z_i(t) | 0 \leq \beta \leq 1 \wedge t \in \mathbb{R}\} = Y_{ji},$$
$$X_{ij} \subset \mathbb{R}_0^+, Y_{ji} \subset \mathbb{R}_0^+.$$

1.3 The Associate Parameter Family

We introduce the parameter space P,

$$P = \mathbb{R}_0^+ \times \mathbb{R}_0^+ \times \mathbb{R}_0^+ \times \mathbb{R}_0^+ \times E,$$

where

$$E = \{x | x \in \mathbb{R} \wedge 0 \leq x \leq 1\}.$$

For every quintuple $p = \langle k_1, K_1, k_2, K_2, \beta \rangle$, $p \in P$, (A.4) determines a dynamical system $\hat{s}(p)$ with structure

$$\langle \mathbb{R}_0^+, \{\mathbb{R}_0^+ \times \mathbb{R}_0^+, \mathbb{R}_0^+ \times \mathbb{R}_0^+\} \rangle.$$

Then s,

$$s = \bigcup_{p \in P} \hat{s}(p)$$

is clearly an input-output system with the parametrisation $\hat{s}: P \to \mathcal{P}(s)$ of which we list some of the critical values. We first observe that $\wedge_i t(Z_i(t) = 0)$ if $k_i = K_i = \beta = 0$ for some i. For $q_1 = \langle k_1, K_1, 0, 0, 0 \rangle$ with $k_1 > 0$ and $K_1 > 0$, one then has

$$\hat{s}(q_1) \subset (X_{11} \times \{0\}) \times (Y_{11} \times \{0\}),$$

and for $q_2 = \langle 0, 0, k_2, K_2, 0 \rangle$ with $k_2 > 0$ and $K_2 > 0$,

$$\hat{s}(q_2) \subset (\{0\} \times X_{22}) \times (\{0\} \times Y_{22}).$$

For every $p' \in P$, $p' = \langle k_1, K_1, k_2, K_2, 0 \rangle$ with

$$k_1 > 0, K_1 > 0, k_2 > 0, K_2 > 0 \tag{A.5}$$

and

$$\hat{s}(p') \subset (X_{11} \times X_{22}) \times (Y_{11} \times Y_{22}),$$

$\hat{s}(p')$ is a system component of the trivial coupling $\hat{s}(q_1) \otimes \hat{s}(q_2)$. And for $p'' = \langle k_1, K_1, k_2, K_2, \beta \rangle$, with $k_1, \ldots, K_2$ as in (A.5) and $0 \leq \beta \leq 1$, the systems $\hat{s}(q_1)$ and $\hat{s}(q_2)$ are the decoupled subsystems of $\hat{s}(p'')$, meaning that the coupled subsystems of $\hat{s}(p'')$ go over into $\hat{s}(q_1)$ and $\hat{s}(q_2)$ as β approaches zero.

This example of ecological interaction between species confirms the rather trivial experience—incomprehensible to holists, however—that systems may indeed change their behaviour or other attributes depending on the nature and strength of their bonds with other systems, but that such changes may well be state-determined effects.

1.4 State Determinacy and Hierarchical Evolution

Finally, we employ the parametrisation $[\hat{s}(p)]_{p\in P}$ of s to characterise evolutionary processes and stratified systems. Considering the discussion of (III.26), the family $[\hat{s}(p)]_{p\in P}$ not only covers the state-determined subrelations of s; it also determines all the modes of interaction which may arise between any two systems contained in it, depending solely on the values of β.

Again assume the case (A.5) with fixed k_1, K_1, k_2, K_2. For every x, $0 \leq x \leq 1$, define the interval

$$I_{0x} = \{t \mid 0 \leq t \leq x\}.$$

If $x > 0$, then

$$\pi(I_{0x}) = \{\langle p, \hat{s}(p)\rangle \mid p \in P \wedge p = \langle k_1, K_1, k_2, K_2, \beta\rangle \wedge \beta \in I_{0x}\}$$

is an evolutionary process with respect to hierarchical complexity. For $\hat{s}(p)$ has zero coupling degree in $[\hat{s}(p)]_{p\in P}$ when $\beta = 0$, whereas its coupling degree equals 1 for $\beta > 0$. Moreover, there is a state-determined hierarchy of coupling relations based on $[\hat{s}(p)]_{p\in P}$ whose number of strata increases from 1 to 2 in the course of $\pi(I_{0x})$.

Part Two
The Evolution of Social Structure

IV Perspectives on Non-Adaptive Evolution

1 Is Sociobiology Reductionist?

In the biobehavioural sciences, explanations of animal social behaviour constitute a class of theoretical problems of their own. To be sure, ethology and behavioural ecology in the traditions of Lorenz, Tinbergen and other have been extremely successful in demonstrating close adaptive relationships between species-specific behavioural traits and environmental features (Eibl-Eibesfeldt 1980): When properly conceptualised, organismic movements, postures and emotional expressions can be treated within the same Darwinian theoretical framework as anatomical characters, namely phylogenetic descent and modification through natural selection, even though behavioural traits are distinct from anatomical ones by individual flexibility, momentary expression and incapacity to fossilise. The methods and findings of classical ethology also prove largely compatible with, and susceptible to, those of other life sciences such as endocrinology, physiology, and functional and developmental biology (Hinde 1982).

1.1 Complexities of Social Interaction

But apart from the intrinsic methodological problems of observing, describing and classifying behaviour (Martin and Bateson 1986), the adaptive significance of a particular social action pattern is in itself hard to specify. In social interactions, the Darwinian fitness of a given phenotype depends critically on how other conspecifics are behaving, that is, the selection of behavioural phenotypes depends on their frequencies in the population (Maynard Smith 1982, p. 1). One of the implications of this frequency dependence is that social behaviour patterns tend to co-evolve under the action of natural selection. Co-evolving traits are sometimes said to form an *adaptive complex*, meaning that the adaptive significance of one character cannot be determined independently of that of the others. The co-evolutionary effects inherent in social behaviour are particularly well documented and discussed in detail in Trivers' (1985) recent book on social evolution.

We illustrate the frequency-dependent selection of social behaviour by contrasting it with a classical example from ethology. By *ritualisation* ethologists understand the adaptive evolution of previously non-communicative traits into species-specific signals. Often ritualisation proceeds so as to adapt display movements to pre-existing receptor organs and physiological releasing mechanisms (Wilson 1975, p. 224). In such cases, increases in fitness arise from the enhanced perceptibility, distinctness and emotive capacity of the signal. These properties, in turn, depend on the physical features of the environment and the receiver's cognitive physiology (Krebs and Davies

1981, Chap 11). Selection is governed by physical and somatic factors, and independent of display frequencies. The appropriate analytic approach is optimisation theory, especially mathematical decision theory (McCleery 1978; Krebs and Davies 1981, Chaps. 3, 11). In the ritualisation of social behaviour (e.g., dominance-submission displays), however, selection proceeds so as to adapt one action pattern to another one which itself is changing in frequency under natural selection. The relevant analytic approach is game theory, leading to complicated non-linear mathematics which constitutes the subject matter of the following two chapters.

Further theoretical problems posed by the Darwinian evolution of social behaviour grow out of the questions of how much of the observed organic variation is adaptive, and how much of the observed similarity is analogous rather than homologous. As for behavioural ecology, in general, comparative ethologists have been quite successful in designing methods to increase the confidence of adaptive evolutionary hypotheses and explanations. These methods include multivariate factor analyses and statistical correlations between environmental constraints and the phenotypic variation observed not only within and between species, but also between different taxonomic levels (Wilson 1975, pp. 551–552; Krebs and Davies 1981, Chap. 2; Ridley 1983). As for social behaviour in particular, the social environment of a given phenotype is part of the variation from which behavioural adaptations are selected, rather than providing independent constraints on adaptation. Hence the inherent difficulty to trace the relevant hereditary variation on which the natural selection of social behaviour operates becomes apparent.

Eventually, since Darwin's day, an apparent paradox of co-operative behaviour has been considered the most serious challenge to evolutionary, adaptive explanations of animal social structure. A behavioural phenotype is said to be *co-operative* if an organism expressing it engenders fitness benefits to others than his immediate descendants (e.g., alarm calls in birds). Now co-operative actions which in addition reduce their performers' own fitness ("altruism") characterise many social animal species, especially the eusocial insects with their sterile workers castes. But altruistic restraints on reproductive behaviour cannot evolve by means of natural selection acting among individual organisms (Darwin 1859, pp. 257–262; Wilson 1975, Chap. 1; Wittenberger 1981, Chaps. 3, 4). Invoking concepts of group selection, evolutionary biologists often tried to resolve this paradox by associating altruism with group benefits. However, Williams (1974, Chap. 4), Maynard Smith (1976) and others provided strong arguments in favour of the thesis that under most realistic circumstances group selection is slow and ineffective as compared to Darwinian selection *sensu stricto*.

1.2 Sociobiology: Merits and Limits

Recent development in the biological study of social behaviour, which field has been named *sociobiology* by E.O. Wilson (1971, 1975), has made the evolution of social structure a major focus of scientific as well as philosophical interest and controversy. One of the most controversial issues is whether there exists a universal biological basis of *all* social behaviour, including that of man at all levels of sociocultural organisation (Wilson 1975, Chap. 1), or whether the pretension of such a basis is guided by certain sociobiologists' "vaulting ambition" rather than well-founded theoretical considerations (e.g., Kitcher 1985). Unfortunately, the issue of the common basis of all co-operative phenomena studied in behavioural science sometimes distracts scientific debates from the more central points made by sociobiologists. Above all, the recent establishment of sociobiology as a biobehavioural subdiscipline of its own must be understood in terms of the introduction, analysis and routine test of a number of novel concepts and theories in behavioural ecology; these concepts and theories have proved unparalleled in resolving the methodological problems and theoretical paradoxes posed by the Darwinian evolution of social behaviour. Introducing a modified concept of genetic fitess ("inclusive fitness"), Hamilton (1964) succeeded in showing that altruism is an adaptive social trait provided genotypes rather than individual phenotypes constitute the units of natural selection. Since the genetic evolution of altruistic behaviour works most effectively among individuals who are closely related by common descent, this mode of behavioural evolution has been dubbed "kin selection" by Maynard Smith (1964). Alternatively, the hereditary basis of behavioural phenotypes can be characterised in the first approximation by the assumptions that there is non-zero genetic variance in these phenotypes and that the individuals expressing them breed true. On these assumptions, the effects of frequency-dependent selection at the phenotypic level can be adequately explained in terms of *individual* fitness, with the use of game-theoretic methods. This approach also offers a number of conceptual and technical advantages. With evolutionary game theory, group selection theory need not be employed to explain ritualised intraspecific competition (Maynard Smith and Price 1973). Specifications of the hereditary variation from which behavioural adaptations are selected become possible in a natural and straightforward fashion (Maynard Smith 1982, pp. 5–8). When combined with the explicit population-genetic approach, evolutionary game theory renders possible useful analyses of correlation effects between genic and phenotypic selection (e.g., Bomze et al. 1983; Thomas 1985a, b).

However, despite their effective conceptual innovations, the sociobiological theories leave fundamental questions of social organisation unresolved. By these questions we do not simply mean issues exposed to ongoing debate, for example, whether natural selection is primarily genic or organismic. Such issues are of empirical nature and can, in principle, be discussed and sooner or later settled on the basis of observational evidence. We rather mean problems posed by the processes of hierarchical differentiation, which deserve special attention in the present context for at least two reasons. Firstly, hierarchical differentiation must generally be viewed as a mode of non-adaptive evolution. This view constitutes the subject matter of the following chapters. Secondly, serious criticisms of sociobiology refer to the hierarchical nature of biological organisation. They imply the irreducibility of (theories of) hierarchical complexity, while rejecting the allegedly reductionist sociobiological explanations of this complexity. Their proponents claim that they are of fundamental conceptual, as compared to empirical, significance. They shall now be considered with reference to the results of Part One.

1.3 The Quest for Alternatives

In the sociobiology debate of the present decade, charges of reductionism usually refer to one of two arguments (if not to both at the same time). Authors such as G.C. Williams, E.O. Wilson and R. Dawkins emphasised the notion that the gene selectionist concepts of "selfish" DNA and inclusive fitness are indispensable for the biological explanation of social behaviour. Now this notion has been accused of entirely neglecting the environmental constraints on the ontogenetic development of behavioural phenotypes, and has accordingly been termed "genetic determinism". For a general review and discussion of the issue of genetic determinism, we refer to Dawkins (1982).

Evolutionary biologist S.J. Gould (1977, 1980, 1982) has put the issue into the context of multilevel organisation in biological systems. Gould believes that genetic determinism cannot be adequately debated merely as a factual problem of possible alternative units of natural selection. In his opinion, natural selection and adaptation are inherently multilevel processes working simultaneously on DNA, organisms, demes and species. Gould's proposition of a more general, multilevel evolutionary process must thus be taken as the utmost antithesis to genetic reductionism.

The second major critique has been phrased as an argument against the "adaptationist programme" of sociobiology (Lewontin 1979; Gould and Lewontin 1979). This argument is even broader in scope than that against genetic determinism because it also aims at those sociobiological theories which dispense with explicit reference to the genetic basis of social behaviour (e.g., game theory as applied to the selection of phenotypes, or R.D. Alexander's (1979) account of human sociobiology). For Gould and Lewontin, "adaptationism" has many connotations. Here, we concentrate on the allegedly reductionist ones. Studies under the adaptationist programme are asserted to proceed in two steps: Firstly, an organism is "atomised" into parts (usually referred to as "traits"), each being explained as a structure optimally designed by natural selection for its function. But organisms are "integrated wholes", not collections of discrete objects. Hence optimality theory as applied to the functions of the parts is misleading; examples are furnished. Secondly,

"after the failure of part-by-part optimisation, interaction is acknowledged via the dictum that an organism cannot optimise each part without imposing expenses on others. The notion of 'trade-off' is introduced, and organisms are interpreted as best compromises among competing demands. Thus, interaction among parts is retained completely within the adaptationist programme..." (Gould and Lewontin 1979, p. 585).

It is concluded that adaptationism cannot work because it is so firmly based on the reductionist premise of wholes being no more than the sums of their parts.

A formidable collection of genetic determinisms, adaptationisms, and related reductionisms has been sampled from the Chamber of Horrors of -*isms* by Kitcher (1985) in his quest for sociobiologism. However, the results of Part One suggest that anti-reductionist arguments of the holistic type cannot be seriously raised against sociobiological theories. Even in those instances in which the latter are truely reductionist (see below), nothing could be advanced against them except that their axioms were logically inconsistent, or their consequences would contradict observation. Sociobiology, as most other biological disciplines, is indeed concerned with the hierarchical organisation of living systems. But theories of higher-level social interactions may be transformed into ones of lower-level structures in such a way that their material contents are preserved. This possibility has been explicitly demonstrated for first-order theories and their semantic structures in Section III.3. Conversely, the fact that hierarchical complexity is generally not an adaptive feature of organic systems is empirically contingent and certainly non-analytic in any of the available theories of hierarchical organisation.

As an example, consider Hamilton's account of kin altruism "in a world of model organisms whose behaviour is determined strictly by genotype" (Hamilton 1964, p. 16). To make the example even more distinct, add Dawkin's (1976) conception of organisms being biomolecular survival machines for their genes. One thus gets an approach to social behaviour which refers exclusively to interactions at the biochemical level of organisation. Under which conditions could the account be accepted as an empirically reasonable hypothesis? The following conditions are well known to biologists to be necessary. Firstly, there must exist clear one-to-one correspondences between the Mendelian genes postulated by Hamilton and the base-pair sequences in DNA molecules. Secondly, there must be a strict separation between germ line and soma (Weismann's principle) in order that genetic evolution might proceed independently of development. The allelic effects on genotypic fitness must be additive, and the environment must be sufficiently constant and homogeneous if Hamilton's population genetic recursions should hold. Numerous other assumptions concerning the genetic system (linkage, dominance, etc.) must also be made. Now the point is that except for Weismann's principle, which for many species is believed to be exact, these assumptions do not generally hold empirically, but nonetheless provide useful approximations (Maynard Smith 1982, p. 6; Sober 1984a, esp. p. 313). One cannot even reasonably claim that the genetic determinism in the example is at *any* rate oversimplifying. Whether a theory proposes an oversimplified view of the empirical world can only be decided with reference to its intended applications. If, for instance, Hamilton's genetic explanation of social adaptations is applied to the gross patterns of kin altruism in the haplo-diploid social insects, it is of great heuristic value since it provides information about the basic evolutionary processes involved. If, on the other hand, details of the mating systems in these insects are also taken into consideration, the purely genetic explanation may indeed turn out to be oversimplifying when tested against the observational data (Wilson 1971, pp. 329–330; Wittenberger 1981, pp. 529–532). The example, illustrates quite well why evolutionary biologists such as Maynard Smith (1982, p. 6) or Dawkins (1982, Chap. 2, p. 113) refuse to continue debating the issues of genetic reductionism and adaptationism, and suggest to focus attention on the *factual* ranges and limits of the adaptive hypotheses instead.

The crucial problem indeed seems to be that biologists lack a general theory of non-adaptive evolution. To be sure, contemporary biology makes considerable effort towards such various theories. For almost two decades, Gould and his co-workers have been advocating a view of macroevolution which they hope will develop into a general theory of multilevel evolution, extending as well as transcending neo-Darwinism. Other conceptual innovations may emerge from non-equilibrium thermodynamics and the biophysics of entropy flows which have established broad theoretical approaches to natural self-organisation in recent years. However successful these approaches may prove in more narrowly circumscribed biological applications, such as the hypercyclic evolution of the genetic code (Eigen and Schuster 1979; Küppers 1983; Babloyantz 1986), it must be kept in mind that a general theory of natural self-organisation is presently *in statu nascendi* at best, and has had little impact on sociobiology thus far.

The lack of general, empirically satisfying alternatives to the adaptive hypotheses of sociobiology leads to situations like this: In their critique of genetic determinism, the Sociobiology Study Group (1977) accuses sociobiologists of telling "just-so" stories about social adaptations, and they suspect that "this sort of fantasy is likely to become a popular pastime in the near future. We may even envision a parlor game, 'Find the Adaptation'" (p. 146). More recently, Kitcher (1985) resumed this kind of criticism, constantly inventing ad hoc explanations alternative to the general theory of social adaptations. Readers of Kitcher's book may get the impression of a popular anti-adaptationist parlour game, "Find the Alternative".

1.4 An Approach to Non-Adaptive Change

As for Part Two of this book and its perspectives on the issues raised thus far, we do not intend to develop theoretical alternatives to the neo-Darwinian fundamentals of sociobiology. Nor do we pretend to be capable of presenting a unified theory of biosocial evolution, although the following chapters aim at a systematic account of basic evolutionary features that are common to biological *and* cultural systems. What we can do, however, is to treat *selected aspects* of biosocial evolution of which we suggest that they should be covered by any attempt at such a unified theory. The significance of our treatment rests upon the fact that it is more inclusive than the neo-Darwinian approach, yet logically fully consistent with the latter, and applicable to processes of hierarchical differentiation. The analytic framework suitable for such a treatment is parametrisation of laws as described in abstract logical and semantic terms in Part One. We start from the familiar mathematical representations of selection in continuous time that have been widely used by theoretical ecologists, population geneticists, and evolutionary game theorists. Put in technical terms, we take the systems of non-linear differential equations of continuous-time population dynamics as our basic laws. However, we do not specifically look at the dynamics of these systems. We rather treat the coefficients (fitness values, interaction coefficients) entering the basic equations as independent variables, that is, control parameters in the sense of Part One (esp. Appendix). Allowing for non-dynamic variables in population-dynamical systems amounts exactly to what has been called parametrisation of systems in Part One. We then investigate how these systems transform under parameter changes, referring to the transformation as *non-adaptive* (*non-dynamic, secular*) evolution. The typical time scales of secular evolution in our examples are inverse genetic mutation rates and geological time scales. The idea behind this approach is to use one and the same kind of laws (basic equations) to describe modes of adaptive as well as non-adaptive evolution. The examples and applications of the approach are mostly drawn from current sociobiological debates. Hence the claim of this book to contribute to a unified theory of biosocial evolution. Special emphasis will be given to problems of hierarchical organisation in the chapter on human social structure.

2 The Concepts of Structural and Evolutionary Instability

In this chapter, we establish linkages between the conceptual framework of system evolution introduced in Part One, and the game-theoretic approach to animal behaviour. We consider parameter families of systems of interacting behavioural pheotypes and ask how the biologically significant properties of these systems may change under parameter variation. The biological significance of parameter changes, in turn, will be pointed out in the following chapters with reference to various applications of the present formalism.

2.1 The Meaning of Structural Instability

The relevant mathematical approach is structural-stability theory of dynamical systems (Hirsch and Smale 1974; Thom 1975; Mesarović and Takahara 1975, p. 163). Intuitively, the concepts of structural stability and instability of dynamical systems, respectively, mean continuous and discontinuous deformation of phase-space trajectories (cf. Part One, Appendix) under parameter variation. Discontinuously changing phase-space trajectories are also said to *bifurcate*. The concept of structural *in*stability, or bifurcation, is particularly relevant to the subject matter of the present inquiry. Theoretical population biologists often exclude structurally unstable systems from their analyses. They argue that the mathematical approach to population dynamics can only yield approximate descriptions of the reality so that a parameter perturbation of a given dynamical system may well be as good as the system itself. They conclude that from the point of view of applications only properties of dynamical systems are meaningful which persist under parameter perturbation since they are the only properties reliable for test and prediction. From an evolutionary point of view, however, structural instabilities indicate fundamental transitions in the structure and complexity of population-dynamical systems subject to changing constraints. In particular, increases in *hierarchical* complexity may correspond to structural instabilities, as the example in the Appendix (Part One) shows. In this example, previously decoupled populations enter into interaction and change their dynamic behaviour qualitatively when the interaction parameter increases from zero to some positive value.

However, we restrict our explicit definitions of the concepts of structural stability and instability to the equilibrium properties of dynamical systems rather than to the dynamics in general. The reason for this limitation is that we concentrate on long-term evolutionary phenomena which in many of our examples correspond to stationary states and asymptotic stability. Thereby the various concepts of dynamic (as compared to structural) stability, including asymptotic stability, are supposed to be well known (for reference, see Hirsch and Smale 1974, Chap. 9). Occasionally, when the dynamical systems in point may not admit asymptotically stable equilibria, we may also consider the structural stability and instability of time averages of dynamic variables (Sect. V. 2). The following definitions are then tacitly supposed to be modified in appropriate ways. Adopting the mathematical terminology of dynamical systems theory, we also use the term "attractor" synonymously with "asymptotically stable equilibrium".
For $k \geq 1$, $m \geq 1$, let

$$\frac{dX_i}{dt} = F_i(C_1,\ldots,C_k; X_1,\ldots,X_m), \quad 1 \leq i \leq m, \tag{IV.1}$$

be a system of dynamic differential equations in continuous time. The functions F_i depend on the real-valued dynamic variables $X_1,\ldots,X_m$ and the control parameters

$C_1, \ldots, C_k$, but do not depend explicitly on time t ("autonomous" dynamical system). Let

$$F_i(C_1, \ldots, C_k; \bar{X}_1, \ldots, \bar{X}_m) = 0, \quad 1 \leq i \leq m, \tag{IV.2}$$

with the bars indicating equilibrium values. The equilibrium $\bar{X} = \langle \bar{X}_1, \ldots, \bar{X}_m \rangle$ is called *structurally stable* if and only if for every sufficiently small perturbation c_j of C_j ($1 \leq j \leq k$), there exists a unique perturbation x_i of $\bar{X}_i$ so that $\bar{X} + x = \langle \bar{X}_1 + x_1, \ldots, \bar{X}_m + x_m \rangle$ is itself an equilibrium,

$$F_i(C_1 + c_1, \ldots, C_k + c_k; \bar{X}_1 + x_1, \ldots, \bar{X}_m + x_m) = 0, \quad 1 \leq i \leq m, \tag{IV.3}$$

with the same dynamic stability properties as $\bar{X}$. The equilibrium $\bar{X}$ is *structurally unstable* if and only if it is not structurally stable. Note the difference between *dynamic* and *structural stability*. "Dynamic stability" refers to system responses to perturbations of the dynamic variables, whereas "structural stability" refers to system responses to parameter perturbations.

As an example, let $\bar{X}$ be asymptotically stable, meaning that there exists a neighbourhood $U(\bar{X})$, $U(\bar{X}) \subset \mathbb{R}^m$, in which every solution of (IV.1) converges to $\bar{X}$. Then the structural stability of $\bar{X}$ implies that if $x \in \mathbb{R}^m$ satisfies (IV.3), $\bar{X} + x$ is asymptotically stable, too.

2.2 Structurally Stable and Unstable Games

We proceed to link the concepts of structural stability and instability to the framework of evolutionary stability/instability, which is of fundamental importance for the biobehavioural approach to animal social structure. As for the game-theoretic and game-dynamic background of the analysis, we refer to Maynard Smith (1982) and Hofbauer nd Sigmund (1984).

Consider a set of m behavioural alternatives, or *component strategies*, of social interaction available to the members of an animal population. Let $P = \langle P_1, \ldots, P_m \rangle$ be the corresponding frequency distribution of behavioural phenotypes, or *population strategy*, with $P_i \geq 0$ for $1 \leq i \leq m$, and $\sum_{i=1}^{m} P_i = 1$. P is called *mixed* if $P_i \neq 1$ for every i, $1 \leq i \leq m$, and *pure* otherwise. Assume that the matrix $\|C_{ij}\|$, or, equivalently, C gives the pay-off in fitness to animals choosing the component strategy i in interactions with conspecifics of phenotype j ($1 \leq i \leq m$, $1 \leq j \leq m$). Since fitness pay-off is related to the number of offspring, the rate of increase $P_i^{-1} dP_i/dt$ of the phenotype i equals the mean fitness pay-off $(CP)_i$ in excess of the population average $P \cdot CP$. This gives

$$\frac{dP_i}{dt} = P_i(E_i \cdot CP - P \cdot CP), \tag{IV.4}$$

with the unit co-ordinate vectors $E_1, \ldots, E_m$. Obviously, (IV.4) has the form (IV.1); it has first been obtained by Taylor and Jonker (1978) within the dynamic approach to *symmetric* population games, i.e., games with equally matched opponents (not necessarily implying a symmetric pay-off matrix!). The simplex

$$\mathscr{S}_m = \left\{ \langle P_1, \ldots, P_m \rangle \mid \sum_{l=1}^{m} P_l = 1 \wedge P_1 \geq 0 \wedge \cdots \wedge P_m \geq 0 \right\}$$

is well known to be invariant under (IV.4). $\partial \mathcal{S}_m$ is the boundary of $\mathcal{S}_m$. If $1 \leq j \leq m$, "$\partial^{(j)} \mathcal{S}_m$" denotes the jth face of the boundary with $Y_j = 0$.

An equilibrium $\bar{P}$ of (IV.4) is called an *evolutionarily stable strategy*, or ESS, if and only if

$$\bar{P} \cdot C\bar{P} \geq Q \cdot C\bar{P} \tag{IV.5a}$$

for all solutions $Q(t)$, and if for some Q', $Q' \neq \bar{P}$,

$$\bar{P} \cdot C\bar{P} = Q' \cdot C\bar{P} \quad \text{then} \quad \bar{P} \cdot CQ' > Q' \cdot CQ'. \tag{IV.5b}$$

Equivalently, $\bar{P}$ is an ESS of (IV.4) if and only if

$$\bar{P} \cdot CQ > Q \cdot CQ \tag{IV.5c}$$

for every Q, $Q \neq \bar{P}$, in a neighbourhood $U(\bar{P})$ (Hofbauer et al. 1979). An equilibrium of (IV.4) is called *evolutionarily unstable* if it is not an ESS.

Conditions (IV.5) of evolutionary stability can be understood as a combined optimality and stability condition of systems of interacting behavioural phenotypes. The inequality $\bar{P} \cdot C\bar{P} \geq Q \cdot C\bar{P}$ means that $\bar{P}$ is an optimum population strategy since $\bar{P}$ does at least as well against itself as any other solution Q of (IV.4). If, however, a strategy distribution Q', $Q' \neq \bar{P}$, exists which does equally well against $\bar{P}$ as $\bar{P}$ itself ($Q' \cdot C\bar{P} = \bar{P} \cdot C\bar{P}$), then selection nonetheless prefers $\bar{P}$ to Q' since $\bar{P}$ does better against Q' than Q' itself ($\bar{P} \cdot CQ' > Q' \cdot CQ'$).

Taylor and Jonker (1978) have shown that every ESS of (IV.4) is asymptotically stable and that the converse is not true. Structural-stability analyses of systems of the type (IV.4) have been carried out by Zeeman (1980, 1981) and Thomas and Pohley (1982). Here we show that small but otherwise arbitrary perturbations of C leave the equilibrium $\bar{P}$ invariant up to some well-defined perturbation p if $\bar{P}$ is an ESS.

Theorem 1. If $\bar{P}$ is an ESS of (IV.4) with

$$\bar{P}_i(E_i \cdot C\bar{P} - \bar{P} \cdot C\bar{P}) = 0, \quad 1 \leq i \leq m, \tag{IV.6}$$

then $\bar{P} + p$ is an ESS associated with the pay-off matrix $\| C_{ij} + c_{ij} \|$ for all sufficiently small perturbation coefficients c_{ij}, $|c_{ij}| \geq 0$ ($c \neq 0$), and solutions p of the linearised-perturbation of (IV.6) with $\sum_{j=1}^{m} p_j = 0$ and

$$p_i(E_i \cdot C\bar{P} - \bar{P} \cdot C\bar{P}) + \bar{P}_i(E_i \cdot Cp - p \cdot C\bar{P} - \bar{P} \cdot Cp)$$
$$= -\bar{P}_i(E_i \cdot c\bar{P} - \bar{P} \cdot c\bar{P}). \tag{IV.7}$$

Proof. Choose $Y \in U(\bar{P})$ with

$$\bar{P} \cdot CY - Y \cdot CY > 0, \tag{IV.5'}$$

according to (IV.5c). Then

$$(\bar{P} + p) \cdot (C + c)Y - Y \cdot (C + c)Y > 0 \tag{IV.8}$$

for all sufficiently small values of $|c_{ij}|$ and $|p_i|$, where c and p are connected as in (IV.7) and $c \neq 0$. For every such pair of sufficiently small quantities c and p, there is a neighbourhood $U'(\bar{P} + p)$ so that

$$(\bar{P} + p - Q) \cdot (C + c)Q > 0 \tag{IV.8'}$$

for every $Q \neq \bar{P} + p$, $Q \in U'(\bar{P} + p)$. In fact, if in every such neighbourhood there were some Q', $Q' \neq \bar{P} + p$, so that $(\bar{P} + p - Q') \cdot (C + c) Q' \leq 0$, (IV.5c) could not hold. In order to see this, let $c \to 0$. Then the inequality $(\bar{P} + p - Q') \cdot (C + c) Q' \leq 0$ gives $\bar{P} \cdot CQ' - Q' \cdot CQ' \leq 0$, which contradicts (IV.5c) because in this limit $Q' \neq \bar{P}$. Hence, by (IV.8') all sufficiently small structural perturbations lead to ESS's.

Corollary. The perturbation p is uniquely determined by (IV.7).

Proof. (i) If $\bar{P}$ is pure, then $p = 0$ since otherwise $\bar{P} + p$ is an interior ESS, and $\bar{P}$ a boundary ESS, in a suitably chosen subgame of the perturbed game. An interior ESS (i.e., all strategy components active) is the sole attractor of a game, however (Zeeman 1980). (ii) If $\bar{P}$ is an interior ESS on $\mathscr{S}_m$, one has $E_i \cdot C\bar{P} = \bar{P} \cdot C\bar{P}$ and $p \cdot C\bar{P} = 0$ so that

$$E_i \cdot Cp = - E_i \cdot c\bar{P} + \bar{P} \cdot c\bar{P}, \quad 1 \leq i \leq m \tag{IV.7'}$$

(every solution p of (IV.7') satisfies $\bar{P} \cdot Cp = 0$!). Observe that (IV.4) is invariant under the transformation $\|C_{ij}\| \to \|C_{ij} + V_j\|$, where $\langle V_1, \ldots, V_m \rangle$ is an arbitrary vector. Hence $m - 1$ components $V_1, \ldots, V_{m-1}$ may be utilised to remove all possible degrees of degeneracy from the coefficient matrix on the left-hand side of (IV.7'). Kramer's rule then determines p uniquely, whereby the mth degree of freedom due to V_m may serve for normalising p so that $\sum_{j=1}^{m} p_j = 0$. (iii) If $\bar{P}$ is mixed but $\bar{P} \in \partial \mathscr{S}_m$, then $p_k = 0$ if $\bar{P}_k = 0$ and $E_k \cdot C\bar{P} \neq \bar{P} \cdot C\bar{P}$ for some k, $1 \leq k \leq m$. Alternatively, if $\bar{P}_k = 0$ and $E_k \cdot C\bar{P} = \bar{P} \cdot C\bar{P}$, then again $p_k = 0$ by an obvious argument similar to (i). All other components of p follow as described in (ii).

Remark 1. The converse of Theorem 1 (every non-ESS equilibrium has an unstable structural perturbation) does not generally hold. Schuster et al. (1981, p.4) indeed give an example of a non-ESS attractor that is structurally unstable. But Zeeman (1980, p. 476) has shown that the three-strategy game with

$$C = \begin{Vmatrix} 0 & 6 & -4 \\ -3 & 0 & 5 \\ -1 & 3 & 0 \end{Vmatrix}$$

has a non-ESS attractor at the barycentre $\bar{P} = \langle 1/3, 1/3, 1/3 \rangle$ which is structurally stable.

Remark 2. For $C_{ij} = C_{ji}$, (IV.4) reduces to Fisher's selection equations whose asymptotic equilibria are structurally stable (Akin 1979, p. 69). Theorem 1 shows that in the more general case in which $C_{ij} \neq C_{ji}$ may hold the ESS property, which is stronger than asymptotic stability, is sufficient for the structural stability of an equilibrium.

2.3 Informal Summary and Discussion

In a sense, this chapter contains the basic results of Part Two. The game-theoretic account of animal social behaviour has been shown to fit into the conceptual framework of system evolution developed in Part One. The population games of evolutionary game theory can be represented as systems (in the narrow, technical sense of the present system concept) of interacting behavioural phenotypes. Since the properties of population games vary with fitness pay-off, games different by pay-off matrix may constitute parameter families of systems accessible to a unified theoretical treatment. In order to describe phenomena of behavioural evolution governed by variations in fitness pay-off, we have adopted the concepts of structural stability/instability from dynamical-systems theory, and evolutionary stability/instability from evolutionary game theory. "Structural stability/instability" essentially means invariance/change of the characteristics of a given system under variation of its constraints. The term "evolutionary stability" refers to a phenotype distribution in a population which is maintained under selection against any alternative population phenotype. Accordingly, evolutionary instability implies the susceptibility of a population to invasion by selectively more favourable phenotypes. Now the relevance of these concepts to a unifying theoretical treatment of different types of population games, and evolutionary transitions between them, rests upon the following result: If equilibrium population phenotypes are evolutionarily stable, they are structurally stable against perturbations in fitness pay-off.

The latter result has important theoretical implications. Within the class of evolutionary games to which the present analysis is confined (autonomous, frequency-dependent selection equations in continuous time), the concepts of evolutionary and structural stability practically coincide. Referring to this coincidence, we shall use "structural stability/instability" and "evolutionary stability/instability" as synonyma even in contexts in which no explicit reference to game theory is made. The relationship between the concepts of ESS and structural stability established by Theorem 1 and its Corollary enables sociobiology to be explicitly drawn into the evolutionary synthesis of Chapter III – at least to the extent to which sociobiological theories are based on evolutionary game theory. In particular, the concept of ESS, which Maynard Smith and his co-workers have originally associated with the theory of adaptive evolution, turns out to be much broader in content and applicability than its "adaptationist" connotation would suggest. We shall both exploit and extend these results in the following chapters.

V Structural Instability in Evolutionary Population Biology

1 Sociobiology and the Structural Instability of Behaviour Patterns

1.1 Sources of Evolutionary Change

In recent applications of population-genetic and game-theoretic concepts to behavioural ecology, which have become known as "sociobiology", theoretical problems arise from the fact that neo-Darwinian evolution involves processes such as mutation and natural selection which proceed at quite different rates. These problems can be circumscribed as follows. On the one hand, gene selection can in principle be described by dynamic differential equations of the form

$$\frac{dX_i}{dt} = G_i(C; X_1, \ldots, X_m), \quad 1 \leq i \leq m, \tag{V.1}$$

covering the classical selection equations of Fisher, Haldane and Wright. This approach assumes m alleles $A_1, \ldots, A_m$ at certain loci, with respective numbers $X_1, \ldots, X_m$ of copies in the population, whereby G_i is in general a highly non-linear function of $X_1, \ldots, X_m$, of the fitness matrix C, and of epistatic and other effects (Akin 1979). Theoretical interest concentrates on the possibility of asymptotically stable stationary states, or equilibria, of (V.1) with

$$G_i = 0, \quad 1 \leq i \leq m, \tag{V.2}$$

because under certain conditions these equilibria correspond to local maxima of the population mean fitness towards which the gene pool will evolve. Dynamic equations of the type (V.1) have also been used to analyse frequency-dependent selection at the phenotypic level, especially selection of behavioural strategies of intraspecific co-operation and competition. In this approach, which is known as "evolutionary game dynamics" (Taylor and Jonker 1978; Schuster et al. 1981; Zeeman 1981; Hofbauer and Sigmund 1984), $X_1, \ldots, X_m$ are the respective numbers of competitors in an m-strategy game, and C is the pay-off matrix of the game evaluated in terms of Darwinian fitness. Theoretical interest concentrates on equilibria (V.2) that correspond to evolutionarily stable strategies, or ESS's, i.e., population equilibria that are selectively preferred to any alternative strategy distribution (see Sect. IV.2.2).

On the other hand, organic evolution does not mean only dynamic change from disequilibrium population states to stable equilibria if any. Spontaneous mutations at the biomolecular level may at any time introduce additional degrees of freedom

$$G_k \not\equiv 0, \quad k > m, \tag{V.3}$$

into the system (V.1) and cause overall changes in the frequency profiles corresponding to (V.2) if the mutant genes are favoured by selection. Such biochemically induced, non-dynamic transformations of (V.1) characterise both the genetic and the strategic case since genetic mutations may also give rise to novel strategies. In fact, stochastic approaches to population genetics have been developed which explicitly incorporate biomolecular effects ("molecular population genetics"; see Ewens 1979; Nei 1987). But these approaches involve considerable mathematical difficulties and, for the present, seem to have no impact on the solution of those questions addressed below.

Evolution induced by imperfection in biomolecular copying fidelity restricts the theoretical significance of evolutionary analyses based on the dynamic equations (V.1). This may well be illustrated by the notions of ESS and inclusive fitness which are widely accepted as the conceptual fundamentals of the evolutionary approach to animal social structure. Both concepts have been developed within the mathematical framework of (V.1) and the adjoint discrete-generations problem based on non-linear difference equations (Hamilton 1964; Taylor and Jonker 1978; Eshel 1982). According to the general population-genetic explanation scheme, the selected social patterns in animal species are supposed to correspond to genetic dispositions which confer relative reproductive advantage to their individual carriers and thus dominate the asymptotic behaviour of systems of the form (V.1). To explain the independent evolution of similar, functionally analogous social traits in different species, sociobiologists must thus assume that sufficient genetic variation for these traits exists in each of the species in order for selection to be effective. The existence of different, species-specific genetic structures that code for similar phenotypic responses to similar selection pressures, poses a non-trivial theoretical problem, however (Lewontin 1979). The hereditary variation on which selection operates ultimately arises from genetic mutations which geneticists assume to be rare and random with respect to function. Accordingly, Lewontin (1979) insists on the point that biologists concerned with the explanation of independent, analogous adaptations must also answer the question of the availability of the relevant mutational variation. In fact, the adaptive evolutionary explanations of animal co-operative behaviour presuppose, rather than account for, the occurrence of specific, functionally analogous mutations in the genomes of quite different species. However heuristically useful this presupposition may be, it is anything but obvious and ignores the question of why the patterns of kin altruism and ritualised intraspecific competition, with which sociobiology is largely concerned, prove reasonably invariant despite apparently extreme differences in the underlying genetic structures and environmental characteristics.

The following analysis aims at a characterisation, in mathematical terms, of analogous behavioural phenotypes that evolved independently in different species. We look for instability, or evolution, criteria that refer to the branching-off of functions G_k and solutions X_k, $m + 1 \leq k \leq n$, from a finite number m of equations of the type (V.1) so that effects of spontaneous mutations can be encompassed (Allen 1976). Since this amounts to analysing the instability of (V.1) with regard to an unknown number $n - m$ of potential mutant types, this task is difficult to tackle in general (as for the genetic case, see Ewens (1979) and the theories with infinitely many alleles therein). It takes a more manageable form and can be treated by perturbation techniques known from structural-instability analyses of dynamical systems, however, if simplifying assumptions such as weak selection are made. As for Eq. (V.1), such structural instabilities may arise from variations of the fitness coefficients, as has been discussed in the literature (Akin 1979; Zeeman 1980, 1981). Here we are concerned with a special kind of structural instability, namely bifurcations corresponding to (V.3), and we shall prove that they, too, can be characterised in terms of perturbations of the fitness matrix C. This proof leads one step beyond the usual analysis of the asymptotic stability of (V.1) against invasion by rare mutants.

The analysis proceeds as follows. We first treat the genetic case, restricting attention to a single locus under weak selection, with random mating and additivity of the effects of diploidy on allelic fitness. It is shown that the functions G_i take the form

$$G_i(C; X_1, \ldots, X_m) = X_i \sum_{j=1}^{m} C_{ij} X_j / N, \quad 1 \leq i \leq m, \tag{V.4}$$

with the total number $N = \sum_{i=1}^{m} X_i$ of alleles at the locus. In this approach, the system (V.1) combines effects of fertility and mortality selection (Nagylaki and Crow 1974) with those of frequency- and density-dependent selection. The matrix C accordingly depends on allele frequencies and population density. Changes in the allele frequencies $P_i = X_i/N$ are given by

$$\frac{dP_i}{dt} = P_i \left(\sum_{j=1}^{m} C_{ij} P_j - \sum_{j,k=1}^{m} C_{jk} P_j P_k \right) = P_i (C_{i.} - C_{..}), \quad 1 \leq i \leq m, \tag{V.5}$$

where subscript dots indicate statistical averages.

We then analyse the instability of the equilibria (V.2) against invasion by rare mutant alleles, with the use of linear perturbation techniques. The existence and stability of such equilibria are also examined for the present selection mechanisms. Since mutations are supposed to be rare when measured on dynamical time scales, the analysis can be based on the following assumptions. The equilibria (V.2) exist and constitute the relevant long-term properties of (V.1). In particular, (V.2) admits physically realistic, asymptotically stable solutions $\bar{X}_i > 0$, $1 \leq i \leq m$. We show that favourable mutations correspond to characteristic parameter perturbations of the equilibria (V.2). This result is applied to problems of animal social behaviour.

Finally, the strategical interpretation of (V.1) – or, equivalently, (V.5) – will be considered. The basic Eqs. (IV.4) of symmetric dynamic evolutionary games are indeed formally equivalent to Eqs. (V.5) for constant fitness matrix C. This equivalence implies that within the limits of the present approach, the secular evolution of both genetic and phenotypic, polymorphic systems of the form (V.5) can be characterised by the same structural-instability criterion. The extension of this criterion to asymmetric evolutionary games is straightforward.

1.2 A Dynamical Approach to Biosocial Genetics

A single-locus analysis with multiple alleles $A_1, \ldots, A_m$ is carried out in which X_i and P_i, respectively, denote the numbers and frequencies of the alleles, and $N = \sum_{j=1}^{m} X_j$ and $P_i = X_i/N$. It is first assumed that the species is purely diploid so that the total number of individuals in the population is $N_0 = \sum_{i,j=1}^{m} N_{ij}$, where N_{ij} is the number of *ordered* genotypes $A_i A_j$. The locus in question is an autosome, and Hardy-Weinberg proportions obtain (see below). Correspondingly,

$$N = 2N_0$$
$$N_{ij} = P_i P_j N_0 = X_i X_j / 2N, \quad 1 \leq i \leq m, \quad 1 \leq j \leq m$$
$$X_i = 2 \sum_{j=1}^{m} N_{ij}.$$

If $D_{ij}(=D_{ji})$ is the probability per unit time of the death of the genotype $A_i A_j$, the time derivative of N_{ij} reads

$$\frac{dN_{ij}}{dt} = N_0 \sum_{k,l=1}^{m} M_{ik,jl} F_{ik,jl} - D_{ij} N_{ij} \tag{V.6}$$

(Nagylaki and Crow 1974). The coefficient $M_{ik,jl}$ gives the number of ordered matings per unit time between genotypes $A_i A_k$ and $A_j A_l$, and $N_0 F_{ik,jl}$ is the average number of offspring from a single $A_i A_k \times A_j A_l$ union. For random mating,

$$M_{ik,jl} = \alpha N_0^{-2} N_{ik} N_{jl}, \quad \alpha = \text{const.} \tag{V.7}$$

(V.6) goes over into

$$\frac{dN_{ij}}{dt} = \alpha N_0^{-1} \sum_{k,l=1}^{m} F_{ik,jl} N_{ik} N_{jl} - D_{ij} N_{ij}. \tag{V.8}$$

We extend (V.8) so as to include other effects that are not due to mating, for instance, competitive encounters and co-operative interactions. These effects give rise to interaction terms of the form $\sum_{k,l=1}^{m} V_{ij,kl} N_{ij} N_{kl}$ familiar from population dynamics, binary encounters being well-mixed, with the density-proportional frequency of an individual's social interactions. This yields

$$\frac{dN_{ij}}{dt} = \alpha N_0^{-1} \sum_{k,l=1}^{m} F_{ik,jl} N_{ik} N_{jl} - D_{ij} N_{ij} + \sum_{k,l=1}^{m} V_{ij,kl} N_{ij} N_{kl}, \tag{V.9}$$

where

$$V_{ij,kl} = V_{ji,kl} = V_{ij,lk}.$$

Following Nagylaki and Crow (1974), we suppose that the fertility of each mating is composed, in an additive fashion, of the population mean birth rate $R_{..}$, and the deviations $R_{ik} - R_{..}$ and $R_{jl} - R_{..}$ of the individual birth rates R_{ik}, R_{jl} from this mean:

$$\alpha F_{ik,jl} = R_{..} + (R_{ik} - R_{..}) + (R_{jl} - R_{..})$$
$$= R_{ik} + R_{jl} - R_{..}, \tag{V.10}$$

with $R_{ik} = R_{ki}$. Since it is biologically not unreasonable and computationally convenient, we assume additive fitness effects similar to (V.10) for the other coefficients, too,

$$D_{ij} = D_{i.} + D_{j.} - D_{..} \tag{V.11a}$$
$$V_{ij,..} = V_{i...} + V_{j...} - V_{....} \tag{V.11b}$$

with the use of

$$S_{ij} = R_{ij} - D_{ij} \quad \text{(V.12a)}$$

$$V_{ij..} = 2W_{ij} \quad \text{(V.12b)}$$

$$C_{ij} = S_{ij} + NW_{ij}, \quad \text{(V.12c)}$$

(V.9) goes over into

$$X_i\left(\frac{dX_j}{dt} - X_j C_{j.}\right) = -X_j\left(\frac{dX_i}{dt} - X_i C_{i.}\right), \quad 1 \le i \le m, \quad 1 \le j \le m, \quad \text{(V.13)}$$

which can be further decoupled since for $i = j$ one gets

$$\frac{dX_i}{dt} = X_i C_{i.} = X_i(S_{i.} + NW_{i.}) \quad 1 \le i \le m \quad \text{(V.14a)}$$

$$\frac{dN}{dt} = NC_{..} = N(S_{..} + NW_{..}). \quad \text{(V.14b)}$$

Changes in allele frequencies are given by

$$\frac{dP_i}{dt} = P_i(C_{i.} - C_{..}) = P_i(S_{i.} + NW_{i.} - S_{..} - NW_{..}). \quad \text{(V.14c)}$$

The terms $S_{i.}$ and $S_{..}$ entering (V.14) describe the "Malthusian" behaviour of the system. The matrix $\|W_{ij}\|$ specifies the average contribution one individual adds to the reproduction probability per unit time of one genotype $A_i A_j$ through social interactions at the phenotypic level.

We now choose the coefficients R_{ij}, D_{ij} and $V_{ij,kl}$ to be constant with respect to time t, space, population density and allele frequencies. Then, by (V.12), W_{ij} is generally frequency-dependent, and so is C_{ij}, but in addition C_{ij} also depends on the population density. For the population growth rate (V.14b) we postulate

$$\frac{\partial C_{..}}{\partial N} = W_{..} < 0. \quad \text{(V.15)}$$

This condition reflects the influence of ecological limiting factors, as is usually assumed in mathematical population dynamics. Observe that the present additive fitness scheme (V.11) guarantees Hardy-Weinberg ratios for the basic equations (V.9). This can be verified immediately by calculating for (V.9) the analogue of Nagylaki and Crow's (1974) condition of Hardy-Weinberg ratios. Once more following Nagylaki and Crow (1974), one shows that (V.11) is invariant under (V.9). In fact, if (V.11) holds for $t = 0$, it does so for every $t > 0$.

Charlesworth (1980) has pointed out that, when selection is slow, the analogue of (V.14c) for discrete generations also holds for haplo-diploidy. Furthermore, it is possible to avoid Nagylaki and Crow's (1974) difficulty to describe effects of sterility by (V.6). From (V.6) indeed follows that for fertilities $F_{ik,jl}$ with the property (V.10) the minimum birth rate of each genotype is half of the population mean. We simulate sterility and similar reproductive restraints by modified mating frequencies f_{ij}, $0 \le f_{ij} \le 1$, and substitute

$$M_{ik,jl} = \alpha N_0^{-2} f_{ik} f_{jl} N_{ik} N_{jl} \quad \text{(V.7')}$$

for (V.7). Evidently, (V.7′) leaves the assumption of Hardy-Weinberg unchanged, nor does it – after an obvious redefinition of R_{ij} – affect the structure of the governing equations.

1.3 Asymptotically Stable Equilibria

We show that in the present single-locus multiallele system with a frequency- and density-dependent fitness matrix, asymptotic stability is obtained if, as a sufficient condition, selection is weak.

Suppose that n is an integer with $n \geq m + 1$, but arbitrary otherwise so that $n - m$ is the number of distinct kinds of alleles which, at the equilibrium (V.2), may arise from the alleles $A_1, \ldots, A_m$ through biomolecular replication errors. With the mutants present and with subscript asterisks indicating averages taken over the frequency distribution $P_1, \ldots, P_n$, the dynamics is given by

$$\frac{dX_i}{dt} = X_i B_{i*}, \quad 1 \leq i \leq n, \tag{V.1′}$$

where B is the extended fitness matrix which contains C as a submatrix in an obvious fashion.

The application of the present formalism to the extended dynamical system (V.1′) gives

$$B_{ij} = \tilde{S}_{ij} + N\tilde{W}_{ij} = \tilde{S}_{ij} + N\tilde{V}_{ij**}/2, \quad 1 \leq i \leq n, \quad 1 \leq j \leq n \tag{V.16}$$

$$\frac{dP_i}{dt} = P_i(B_{i*} - B_{**}) \tag{V.17a}$$

$$\frac{dN}{dt} = NB_{**}, \tag{V.17b}$$

the meaning of $\tilde{S}_{ij}$, $\tilde{W}_{ij}$, etc., being obvious. One verifies easily that

$$\frac{dB_{**}}{dt} = 2 \sum_{i=1}^{n} P_i(B_{i*} - B_{**})^2 + N \sum_{i=1}^{n} P_i(B_{i*} - B_{**})\tilde{V}_{**i*}$$

$$+ NB_{**} \frac{\partial B_{**}}{\partial N}$$

$$= 2 \operatorname*{Var}_{i \leq n}(B_{i*}) + N \operatorname*{Cov}_{i \leq n}(B_{i*}, \tilde{V}_{**i*}) + NB_{**}\tilde{W}_{**}, \tag{V.18}$$

where the symbols "Var" and "Cov" denote statistical variance and covariance respectively. If selection is sufficiently weak, the variance and covariance in (V.18) are small compared to $NB_{**}\tilde{W}_{**}$. This holds even close to the equilibria of (V.17), where B_{**} is small, too, but $NB_{**}\tilde{W}_{**}$ is still only a first-order small quantity because $\tilde{W}_{**} \neq 0$ is assumed. Hence

$$\frac{dB_{**}}{dt} < 0 \quad \text{if } B_{**} > 0 \tag{V.19a}$$

and

$$\frac{dB_{**}}{dt} > 0 \quad \text{if } B_{**} < 0. \tag{V.19b}$$

We show that every equilibrium $\langle \bar{P}_1, \ldots, \bar{P}_n, \bar{N} \rangle$ of (V.17) with

$$\bar{B}_{i*} = \bar{B}_{**} = \overline{\frac{dB_{**}}{dt}} = 0 \tag{V.20}$$

is asymptotically stable. Because of (V.19) and (V.20) there is a neighbourhood Ω of $\langle \bar{P}_1, \ldots, \bar{P}_n, \bar{N} \rangle$ in which

$$\lim_{t \to \infty} B_{**} = \lim_{t \to \infty} \frac{dB_{**}}{dt} = 0. \tag{V.21}$$

Observe that $P_i(0) > 0$. From (V.17a) it follows that

$$0 = \lim_{t \to \infty} \frac{\ln P_i(t) - \ln P_i(0)}{t} = \lim_{t \to \infty} t^{-1} \left(\int_0^t B_{i*}(T) dT - \int_0^t B_{**}(T) dT \right). \tag{V.22}$$

The last term on the right-hand side of (V.22) vanishes. Then

$$0 = \lim_{t \to \infty} t^{-1} \int_0^t B_{i*}(T) dT = \lim_{t \to \infty} B_{i*}, \quad 1 \leq i \leq n, \tag{V.23}$$

by l'Hôpital's rules, with (V.21) and (V.23) being the desired result.

Because of $\partial B_{**}/\partial N < 0$, the equation $B_{**} = 0$ implicitly defines N as a smooth function of $P_1, \ldots, P_n$ in Ω. However, it is easy to see that with frequency-dependent B_{ij}, $\partial N/\partial P_i = 0$ does *not* generally hold so that equilibria do not locally maximise N as a function of $P_1, \ldots, P_n$ (cf. Slatkin 1979). But if $B_{**} > 0$ for some $\langle P_1, \ldots, P_n \rangle$ in Ω, there is a monotone increase in population size as (V.17) approaches equilibrium.

Similar results are obtained if in (V.16) B_{ij} takes the form of the inclusive fitness (V.33) introduced below, and (V.18) and (V.22) are modified accordingly.

1.4 Favourable Mutations

Now consider the matrix

$$\|\Phi_{ij}\| = \left\|\overline{\frac{\partial G_i}{\partial X_j}}\right\|, \quad 1 \leq i \leq n, \quad 1 \leq j \leq n. \tag{V.24}$$

The horizontal bar indicates the evaluation of the partial derivative at an initial equilibrium of the extended dynamics (V.1), that is, $X_i = \bar{X}_i > 0$ if $i \leq m$, and $\bar{X}_i = 0$ otherwise. Such an equilibrium is unstable against the spread of mutants with initial growth rate ω if the characteristic equation

$$\det \|\Phi_{ij} - \omega \delta_{ij}\| = 0 \tag{V.25}$$

has at least one root ω with a positive real part. Inserting (V.1) into (V.24), one finds

$$\Phi_{ij} = 0 \quad \text{if } m+1 \leq i \leq n, 1 \leq j \leq m, \tag{V.26a}$$

and

$$\Phi_{ij} = \delta_{ij}\bar{B}_{i*} = \delta_{ij}\bar{B}_i. \quad \text{if } m+1 \leq i \leq n \text{ and } m+1 \leq j \leq n. \tag{V.26b}$$

From (V.26a) it follows that the determinant (V.25) reduces to the product of two subdeterminants

$$\det \| \Phi_{ij} - \omega\delta_{ij} \| = \det \| \Phi_{gh} - \omega\delta_{gh} \| \det \| \Phi_{kl} - \omega\delta_{kl} \| = 0,$$
$$1 \leq i \leq n; \quad 1 \leq j \leq n; \quad 1 \leq g \leq m; \quad 1 \leq h \leq m; \tag{V.25'}$$
$$m+1 \leq k \leq n; \quad m+1 \leq l \leq n.$$

This decomposition of (V.25) is a general feature of systems with rates of increase of the form (V.4) (Czaplewski 1973; Allen 1976). The first determinant on the right-hand side of (V.25') characterises the asymptotic stability of the allelic system before the mutants appear. Given (V.26b), the vanishing of the second determinant on the right-hand side is equivalent to $n-m$ decoupled equations for $n-m$ roots of (V.25)

$$\omega_k = \bar{B}_{k*} = \bar{B}_k., \quad m+1 \leq k \leq n. \tag{V.27}$$

Assume now that for every k in (V.27), there is an index i_k, $i_k \leq m$, so that the allele A_k arises from rare mutation events $A_{i_k} \to A_k$. Since selection is supposed to be weak, B_{kj} may be written in terms of a small perturbation b_{kj} of C_{i_kj}:

$$B_{kj} = C_{i_kj} + b_{kj}, \quad 1 \leq j \leq m, \quad m+1 \leq k \leq n.$$

If $i_k \neq i_{k'}$ for every pair k, k' with $k \neq k'$, define

$$c_{ij} = \begin{cases} b_{kj} & \text{if } i = i_k \\ 0 & \text{otherwise} \end{cases} \tag{V.28}$$

where $i \leq m$, $j \leq m$. If two or more mutations with indices $k, k', \ldots (m+1 \leq k \leq n, m+1 \leq k' \leq n, \ldots)$ arise from one and the same allele so that $i_k = i_{k'} = \cdots$, define

$$c'_{ij} = \begin{cases} b_{k'j} & \text{if } i = i_{k'} \\ 0 & \text{otherwise} \end{cases} \tag{V.28'}$$

in addition to (V.28) etc. Accordingly, the following analysis can be restricted to the case $i_k \neq i_{k'}$ for $k \neq k'$ without loss in generality. One gets

$$\omega_k = \bar{C}_{i_k.} + \bar{c}_{i_k.} = \bar{c}_{i_k.}, \tag{V.29}$$

where $\bar{c}_{i_k.}$ is a parameter perturbation of the stationary growth rate $\bar{C}_{i_k.} = 0$.

Equations (V.27) and (V.29) have several important properties. Condition (V.25) governing stability has become completely decoupled. This means that according to (V.27) and (V.29) the instability of the gene pool against simultaneous invasion by several mutant types can be checked for each mutation separately. This result reflects the fact that initially mutants are rare, with negligibly little interference between novel alleles. But in contrast to (V.27), (V.29) depends explicitly neither on the fitness matrices B and C nor on the function G_k specific of the kth type of mutation. With respect to fitness, it depends, in an explicit fashion, solely on the differentials (V.28) which enter the analysis as perturbations of the extant fitness values. One can exploit this dependence of (V.29) on fitness differentials rather than on actual fitness values to characterise the bifurcations (V.3) in terms of parameter perturbations of (V.1):

A one-locus system with equilibrium allele distribution $\bar{X}_1,\ldots,\bar{X}_m$ ($\bar{X}_j > 0$, $1 \leq j \leq m$) is structurally unstable against exactly those mutational steps satisfying

$$G_i(\bar{c}; \bar{X}_1,\ldots,\bar{X}_m) > 0 \tag{V.30}$$

for at least one index i, $1 \leq i \leq m$, where c is the matrix of fitness differences between the mutant and extant alleles.

The significance of (V.30) rests upon the fact that in contrast to (V.27) the condition (V.30) is no longer restricted to selection within systems of specified, resident and mutant alleles (independence of (V.30) of C and B). Within the limits of the present approach, (V.30) rather characterises the selectively admissible *directions* (fitness *differences* c) of mutational change relative to arbitrary ranges of existing genetic variation (arbitrary C; see Fig. V.1). Hence it should prove useful in the genetic

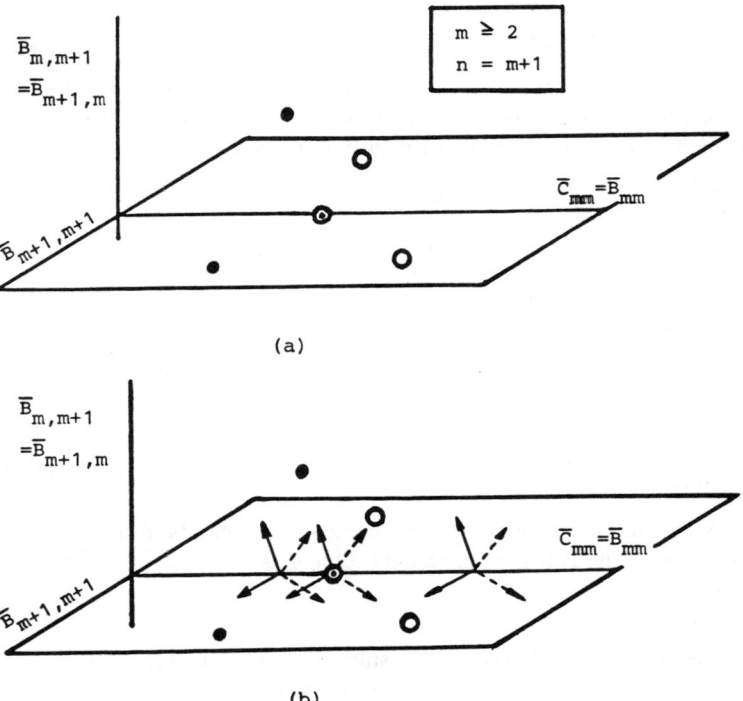

Fig. V.1a,b. Low-dimensional projection of fitness space. The situation depicted illustrates the difference between the selection conditions (V.27) and (V.30). It is assumed that the allele A_{m+1} arises from a mutation $A_m \to A_{m+1}$. **a** The *encircled dot* represents the (m,m)-element of the unperturbed fitness matrix, while the other *solid* and *open* circles indicate fitness matrices after favourable and unfavourable mutations respectively. Whether a point $\bar{B}$ in the fitness space carries a solid or open circle depends on its position relative to ⊙ and is determined by $\omega_k > 0$ or $\omega_k \leq 0$ according to (V.27). **b** Condition (V.30) determines *directions* rather than points in the fitness space. *Solid* and *dashed arrows* correspond to the directions of favourable and unfavourable mutational steps respectively. Since (V.30) is independent of the unperturbed matrix $\bar{C}$, the arrows are independent of ⊙ and can be attached to *any* point on the $\bar{C}_{mm}$-axis. Similar figures can be drawn for the horizontal $\bar{C}_{im}$-axis ($i < m$)

analysis of analogous traits and phenotypic similarities that evolved independently in different species.

In general, the criterion (V.30) says nothing about the asymptotic behaviour of the system with any mutants present. On the assumption of weak selection, however, certain asymptotic properties of the extended dynamics can be inferred from (V.17b) and (V.30). We can show that $B_{**} > 0$ after a mutation $A_{i_k} \to A_k$ for which (V.29) is positive, that is, for which (V.30) holds. Let $\underline{p}_{i_k}$ and p_k be the mutational perturbations of the equilibrium frequencies $\bar{P}_{i_k} > 0$ and $\bar{P}_k = 0$, respectively, so that $p_k = -p_{i_k} > 0$. Using the definition (V.28), we get in linear approximation

$$B_{**} = 2p_{i_k}\bar{C}_{i_k\cdot} + 2p_k\bar{B}_{k\cdot} = 2p_k\bar{c}_{i_k\cdot}. \tag{V.31}$$

Hence (V.31) is positive if $\bar{c}_{i_k\cdot} > 0$. So if just one type of mutation occurs for which (V.30) holds, the mutant allele causes a monotone increase in population size and will be selected for finite asymptotic frequency. If two or more mutant types satisfying (V.30) arise simultaneously this again induces monotone population growth, with at least one (though perhaps not all) of the novel alleles surviving for $t \to \infty$. Stronger results concerning the final population state can be obtained in the game-dynamical interpretation of the formalism (see below) in which the fitness pay-off matrix is constant.

1.5 Application to Inclusive-Fitness Theory

Using (V.30), we show that, in the one-locus approach to the genetics of social behaviour, populations of selfish phenotypes are structurally unstable against the evolution of kin altruism. Thereby we refer to Hamilton's (1964) account in terms of inclusive fitness. Alternative approaches based on exact population genetic recursions have been reviewed and discussed by Boorman and Levitt (1980), Charlesworth (1980), Uyenoyama and Feldman (1980) and, more recently, by Grafen (1985a). But since inclusive-fitness theory combines mathematical simplicity with reasonable quantitative accuracy (Charlesworth 1980; Grafen 1985b), it is more suitable for demonstrating an instance of structural instability of the type (V.30).

According to inclusive-fitness theory, sterility and other reproductive restraints in the social animal species constitute social patterns which, although disadvantageous to personal fitness, are favourable to genotypes governing unselfish behaviour. Altruism works most effectively in kin groups where interacting individuals tend to carry genes identical by descent. If K is the ratio of the partner's gain to the actor's loss in fitness, and r (coefficient of relatedness) gives their expected fraction of genes identical by descent, acts detrimental to the actor's reproduction will increase his inclusive fitness if

$$K > 1/r \tag{V.32}$$

(Hamilton 1964).

The concept of inclusive fitness can be incorporated in the present formalism as follows. Suppose that in pairwise intraspecific interactions donors and recipients can be distinguished and that the categories of donor and recipient are disjoint (donors never act as recipients, and conversely). On the population average, stereotyped social interactions between relatives of degree r change the fitness of a donor of genotype

A_iA_j by the "cost" $-W_{ij}^c(r)$, and that of a recipient by the "benefit" $W_{ij}^b(r)$ per interaction event. Both $W_{ij}^b(r)$ and $W_{ij}^c(r)$ are positive if the interactions are altruistic. Adopting the approximations built into (V.14), one can then follow Charlesworth's (1980) argument and define the inclusive fitness C_{ij} of the genotype A_iA_j by

$$C_{ij} = S_{ij} + \sum_r N(r)(rW_{ij}^b(r) - W_{ij}^c(r)), \tag{V.33}$$

modifying (V.14) accordingly. In (V.33), the expression $rW_{ij}^b(r) - W_{ij}^c(r)$ replaces W_{ij} as the net contribution to the fitness of the genotype A_iA_j. $N(r)$ is twice the average number of an individual's relatives of degree r in the population; its appearance in (V.33) corresponds to the present assumption that the frequency of an individual's social interactions is proportional to density.

Observe that the basic equations (V.14) also hold for the inclusive fitness (V.33). Let $N_{ij}(r)$ be the number of ordered genotypes A_iA_j related by r to some given individual so that

$$N_{ij}(r) = X_i(r)X_j(r)/2N(r), \quad \text{(Hardy-Weinberg)}$$

where $X_i(r) = P_iN(r)$ and $N(r) = \sum_{i=1}^m X_i(r)$. Then the equations preceding (V.14) can be appropriately modified by the kinship interaction term $N_{ij}\sum_r \sum_{k,l=1}^m V_{ij,kl}(r)N_{kl}(r)$. With

$$V_{ij..}(r) = 2(rW_{ij}^b(r) - W_{ij}^c(r)),$$

the calculation of (V.14) proceeds as above.

From (V.33) one has the well-known result that social interactions for which

$$K = \frac{W_{ij}^b(r)}{W_{ij}^c(r)} > 1/r \tag{V.32'}$$

holds increase the inclusive fitness of genotypes controlling kin altruism. Now consider random mutational steps due to replication errors of an allele A_a, where $1 \leq a \leq m$. According to (V.28), the structural variation in fitness resulting from any such step is given by perturbed quantities $S_{ai} + s_{ai}$, $W_{ai}^b(r) + w_{ai}^b(r)$, and $W_{ai}^c(r) + w_{ai}^c(r)$. At equilibrium, the combined average change in fertility and mortality per mutant copy of A_a is $\bar{s}_{a.}$, while $\bar{w}_{a.}^b(r)$ and $\bar{w}_{a.}^c(r)$ measure marginal fitness contributions due to social interactions between relatives. From (V.29) and (V.30) it follows that any such mutant type will spread when rare if and only if

$$\bar{s}_{a.} + \sum_r \bar{N}(r)[r\bar{w}_{a.}^b(r) - \bar{w}_{a.}^c(r)] > 0. \tag{V.34}$$

We first consider mutations affecting behaviour, with no effects on fertility and mortality, i.e., $\bar{s}_{a.} = 0$ in (V.34). Then (V.34) is satisfied if

$$K = \frac{\bar{w}_{a.}^b(r)}{\bar{w}_{a.}^c(r)} > 1/r, \quad 0 < r < 1. \tag{V.35}$$

Although formally identical with (V.32'), (V.35) is different in meaning. It implies that the mutational variation necessary for kin selection will evolve whenever a resident gene gives rise to an irregular copy the fitness increments and decrements which happen to correlate according to (V.35). This result holds for arbitrary populations, including completely selfish and competitive ones, since (V.34) and (V.35) do not depend on the current genotypic interactions and fitness values. In particular, (V.35) shows that the applicability of Hamilton's rule (V.32) is not restricted to the existing ranges of genetic variation, and selection within them, as (V.32′) might suggest. Hamilton's rule equally applies to the selection of the mutational variation for kin altruism in randomly mutating allelic systems (for qualification of this conclusion, see below). In fact, when various types of mutations occur, there is always at least one allele among those mutants satisfying (V.35) which will be selected for non-zero stable frequency, as follows from (V.31). The consequences of these results for certain arguments frequently raised against sociobiology are discussed briefly in the informal summary below.

1.6 Applications to Insect Social Structure

We now consider mutations for which $\bar{s}_{a.} \neq 0$ in (V.34), restricting attention to the limiting case in which donors sacrifice their own reproductive potentials once in their lifetime, but possibly continue providing benefits to others. Sterility among female workers in the social insects and altruistic self-sacrifices can be viewed as instances of this behaviour pattern. Mutational steps towards the evolution of this mode of altruism can be characterized by

$$\bar{w}_{ai}^c(r) = 0, \tag{V.36a}$$

$$\bar{s}_{ai} < 0, \quad 1 \leq i \leq m, \tag{V.36b}$$

meaning that the fitness costs are not attached to individual interactions, but are paid all at once during the ontogenetic development of an altruist. Then (V.36b) implies that sterility may occur among mutant genotypes or their death rates may increase through altruistic self-sacrifices. Let further s_{ai} be decomposed into the respective losses $s_{ai}(r) < 0$ which altruistic phenotypes cause to the fitness of mutant alleles by sacrificing their reproductive capacities on behalf of relatives of degree r:

$$s_{ai} = \sum_r s_{ai}(r) < 0, \quad 1 \leq i \leq m.$$

Then (V.34) and (V.35) go over into

$$\sum_r [\bar{s}_{a.}(r) + r\bar{N}(r)\bar{w}_{a.}^b(r)] > 0, \tag{V.37}$$

$$K = -\frac{\bar{N}(r)\bar{w}_{a.}^b(r)}{\bar{s}_{a.}(r)} > 1/r. \tag{V.38}$$

The important point in (V.38) is that for given $\bar{s}_{a.}(r)$ and $\bar{w}_{a.}^b(r)$ the success of a rare mutant for altruism depends on the number of related beneficiaries. For example, in a diploid species let the mutant genotypes tend to sacrifice their lives to rescue other members of the kin group so that the fitness costs and benefits are about equal, i.e., $-\bar{s}_{a.}(r) \simeq 2\bar{w}_{a.}^b(r)$ (as for the factor 2, observe (V.12b)). The validity of (V.38) then depends on whether $r\bar{N}_0(r) > 1$ holds, where $N_0(r) = N(r)/2$ is the number of kin of relatedness r. This result explicates the common sociobiological notion that altruistic self-sacrifices pay if at least two full sibs, four half-sibs, or eight first cousins, etc. can be saved.

Condition (V.38) suggests interesting applications to the genetic evolution of the large sterile worker castes in the social Hymenoptera. Under certain conditions including haplo-diploidy, hymenopteran female workers tend to be more closely related to each other ($r = 3/4$) than to their mother ($r = 1/2$), the queen of the colony (Hamilton 1964; Wilson 1971; Wittenberger 1981). But as has been emphasised by Hamilton himself, and by others, there must exist mechanisms besides kin selection which contribute to the selective advantage of sterility in the social Hymenoptera if a hymenopteran queen is inseminated by several males during one nuptial flight. In this case, the coefficient of relatedness between the queen and her daughters ranges above that among sisters, and the explanation of sterility on the basis of the condition (V.32) contradicts observation.

However, West (1967) and West Eberhard (1975) pointed out that in semisocial insects the selective advantage of altruism not only depends on effects of genetic relatedness but also on differences in reproductive capacity among co-operating females. Condition (V.38) suggests that this view of kin selection also applies to the evolution of sterility in the eusocial insects. In fact, for a fixed cost-benefit ratio of an altruistic action pattern, (V.38) states that the action maximises the donor's inclusive fitness if he favours relatives for whom $r\bar{N}(r)$ rather than r is maximum. Accordingly, female hymenopteran workers may well be related to their daughters by $r_d = 1/2$, and to their sisters by $r_s \leq 1/2$ if multiple inseminations of hymenopteran queens are the rule. But since the reproductive capacity (size of ovary) of a queen is generally much larger than that of an average female worker (Wilson 1971; Brian 1979), $\bar{N}(r_s) \gg \bar{N}(r_d)$ and $r_s\bar{N}(r_s) > r_d\bar{N}(r_d)$ will hold for workers, and preferring sisters to daughters remains the inclusive fitness maximising trait.

Condition (V.38) suggests a similar argument concerning the queen's control of the reproductive behaviour of female workers (Wilson 1971; Wittenberger 1981). In the haplo-diploid insects, workers are related to their sons by 1/2, and to their brothers by 1/4, which is generally believed to be an important factor in the evolution of the workers' capacity to lay unfertilised eggs from which males develop. Egg-laying by workers is clearly contrary to the reproductive interests of the queen, who should inhibit it. But then the question arises why queen control is more effective in some haplo-diploid species than in others (Trivers and Hare 1976; Wittenberger 1981). Now (V.38) suggests that in queen-worker conflicts on male reproduction there should be a considerable bias in favour of the queen for large, caste-specific differences in egg-laying capacity. Brian's (1979) review of both primitively and highly social wasps, bees and ants provides some evidence for this hypothesis: The caste-specific differences in female body size, ovarian development and size of spermathecae vary roughly with the degree of sociality but tend to vary inversely with the frequency of male reproduction by workers (for exceptions to this rule, see Bourke 1988). The example of the social bees is particularly instructive. Workers usually do not produce any males in honeybees and many other highly eusocial bees, nor are they known to be the exclusive parents of males in any eusocial species, but produce most of the males in certain weakly eusocial bumblebees and some stingless bees (Wilson 1971; Oster and Wilson 1978; Wittenberger 1981; Bourke 1988). However, in bumblebees there are no structural differences between reproductive and assimilative castes, and caste intermediates occur (Wilson 1971; Brian 1979). So in these species $r\bar{N}(r)$ should differ less between the brothers, sons and nephews of a female worker, and the queen control of male reproduction should indeed be less effective.

A less qualitative account of the mutational basis of patterns of kin altruism may be expected from an analysis of the complete structural instability condition

$$\sum_r (\bar{s}_{a.}(r) + \bar{N}(r)[r\bar{w}^b_{a.}(r) - \bar{w}^c_{a.}(r)]) > 0.$$

This analysis, however, is beyond the scope of the present work.

Eventually, a few additional qualifications have to be made. The present analysis is based on numerous simplifying assumptions which, in general, will not hold in real population genetic systems (for detailed discussion see, e.g., Boorman and Levitt 1980). However, since the analysis is primarily conceptual in scope, we consider only its *intrinsic* problems here. First, Hamilton (1964) pointed out that (V.32') is not applicable to the selection of new mutations because sibs might or might not inherit the mutation depending on the point in the germ line of the parent at which it had occurred. In general, a finite number of generations must pass before (V.32'), and *a fortiori* (V.34) and (V.35) can be applied. This point is clearly unfavourable to the establishment of altruistic traits. But Hamilton hastened to add that a mutation might overcome any such initial barrier before it has recurred many times. In fact, if selection is weak, the barrier will be low, and the present structural-instability criteria will describe basic properties of the situation after a few generations. Secondly, the above results are insensitive to the effects of small population size and to violations of the assumption of well-mixed binary encounters (density-proportional interaction terms). Condition (V.30) and its applications do not depend on the *frequency* of the mutant alleles. Hence accidents of random sampling with strong impact on rare mutants in small populations do not affect the structural stability of groups of any size. Finally, one may argue that the present assumptions about density-dependent social interactions do not adequately describe interactions among kin. However, the above results are not restricted to strictly density-proportional interaction terms as long as the *frequency* of social interactions is independent of genotype (let $\beta > 0$ and put N^β, $N^\beta(r)$, etc., instead of N and $N(r)$ in the interaction terms).

1.7 Evolutionary Instability in Secular Time Scales

The above results shall now be applied to problems of evolutionary game theory and the frequency-dependent selection of mutant behavioural phenotypes (cf. Maynard Smith and Price 1973; Maynard Smith 1982). Explicit definitions of the relevant game-theoretic concepts have been given in Section IV.2.2. Since the analysis leading to (V.30) is not restricted to symmetric fitness matrices, (V.30) also applies to dynamic population games, in which C is generally non-symmetric. The evolutionary mechanisms of differential reproduction, diploidy and genetic transmission compatible with the theory of ESS's have been considered in various recent publications (e.g., Bomze et al. 1983; Thomas 1985a, b). These considerations suggest that the ESS may prove relevant under fairly general modes of inheritance, for example, in the diploid single-locus multiallele case for a wide range of dominance patterns. Here we concentrate exclusively on the following problem. The conventional game-theoretic approach to selection acting on phenotypic variation is restricted to systems of *specified* (including specified mutant) strategies; it is not concerned, however, with the responses of such systems to mutations that occur *at random* (Lewontin 1982).

Analogously to the genetic case considered thus far, this problem bears directly on evolutionary explanations of functionally analogous behavioural traits that evolved independently in different species.

The following lemma and theorem establish relationships between the evolutionary stability properties of (IV.4) and mutant strategy components with selection equations branching off from (IV.4). The proofs are based on the tacit assumption that pay-off matrices are non-degenerate in the sense that their rows are pairwise different.

Lemma. Let $\bar{P}$ be an ESS of the system

$$\frac{dP_i}{dt} = P_i(E_i \cdot CP - P \cdot CP), \quad 1 \leq i \leq m, \tag{IV.4}$$

C being constant, with $\bar{P}_j > 0$ for every j, $1 \leq j \leq m$. Assume $n - m$ novel strategy components that are all mutants of one and the same strategy component labelled "i" (i fixed, $i \leq m, n \geq m + 1$). Suppose selection is weak, and let the mutants be initially favoured by selection, i.e., have positive initial growth rates corresponding to (V.27) and (V.30). Then all trajectories of the perturbed game flow to $\partial \mathscr{S}_n$, with one and only one of the mutant components having non-zero asymptotic frequency. This strategy component replaces the ith component.

Proof (by induction on n). Let $n = m + 1$ and, without loss in generality, $i = m$. We proceed in three steps. (i) Assume that there is a solution trajectory of the $(m + 1)$-strategy game which is bounded away from $\partial \mathscr{S}_{m+1}$. The time averages

$$\bar{Y}_j = \lim_{T \to \infty} \frac{1}{T} \int_0^T Y_j(t) dt$$

must satisfy the conditions

$$\bar{Y}_j > 0, \quad 1 \leq j \leq m + 1, \tag{V.39a}$$

$$\sum_{j=1}^{m+1} B_{1j} \bar{Y}_j = \cdots = \sum_{j=1}^{m+1} B_{m+1,j} \bar{Y}_j \tag{V.39b}$$

(Schuster et al. 1981), where B is the pay-off matrix of the $(m + 1)$-strategy game. By a suitable barycentric transformation (Zeeman 1980, p. 488; Hofbauer and Sigmund 1984, p. 171) it is always possible to choose

$$\bar{Y}_j = \tfrac{1}{2} \bar{P}_j \quad \text{for } 1 \leq j \leq m \quad \text{and} \quad \bar{Y}_{m+1} = \tfrac{1}{2}.$$

The matrix B being expressed in terms of C and b, (V.39) gives

$$\sum_{j=1}^{m+1} B_{lj} \bar{Y}_j = \tfrac{1}{2} \bar{C}_{l.} + \tfrac{1}{2}(C_{lm} + b_{l,m+1}), \quad 1 \leq l \leq m,$$

$$= \sum_{j=1}^{m+1} B_{m+1,j} \bar{Y}_j$$

$$= \tfrac{1}{2}(\bar{C}_{m.} + \bar{b}_{m+1,.}) + \tfrac{1}{2}(C_{mm} + b_{m+1,m+1}).$$

Since $\bar{P}$ is an interior equilibrium of the original m-strategy game, $\bar{C}_{l.} = \bar{C}_{m.}$ must hold. One thus has

$$C_{lm} - C_{mm} = \bar{b}_{m+1,.} + b_{m+1,m+1} - b_{l,m+1}, \quad 1 \leq l \leq m. \tag{V.40}$$

Now there exists at least one index l', $l' \leq m$, so that $C_{l'm} \neq C_{mm}$. Otherwise $\bar{P} \cdot C\bar{P} = E_m \cdot C\bar{P}$ and $\bar{P} \cdot CE_m = E_m \cdot CE_m$ would hold, meaning that $\bar{P}$ would not be an ESS. Hence, with $C_{l'm} - C_{mm} \neq 0$, (V.40) cannot hold for sufficiently small perturbations (right-hand side). Therefore (V.39) is false, and all trajectories in the interior of $\mathscr{S}_{m+1}$ flow to $\partial \mathscr{S}_{m+1}$. (ii) Let b and c be still related according to (V.28). Then, by Theorem 1 and its Corollary, the m-strategy game with the matrix $C + c$ has a unique ESS $\bar{P} + p = \langle \bar{P}_1 + p_1, \ldots, \bar{P}_m + p_m \rangle$. Observe that $\bar{P}_j + p_j > 0$ for every j, $j \leq m$, if selection is weak (p small). $\langle \bar{P}_1 + p_1, \ldots, \bar{P}_m + p_m, 0 \rangle$ is not an asymptotically stable equilibrium of the $(m+1)$-strategy game since by assumption it is unstable against the spread of rare mutants of the type $m + 1$. Conversely, a "backward" mutation of type m arising from the strategy component $m + 1$ in the equilibrium $\bar{Q} = \langle \bar{P}_1 + p_1, \ldots, \bar{P}_{m-1} + p_{m-1}, 0, \bar{P}_m + p_m \rangle$ apparently has a negative initial growth rate. Since $\bar{Q}$ is also an ESS on $\partial^{(m)} \mathscr{S}_{m+1}$, it is an attractor of the full $(m+1)$-strategy game. (iii) Assume that $\bar{Q}'$ is another attractor of the game. Then $\bar{Q}'_m \neq 0$ and $\bar{Q}'_{m+1} \neq 0$ must hold, while $\bar{Q}'_{l'} = 0$ for at least one index l', $1 \leq l' \leq m$. The former two conditions mean that $\bar{Q}'$ constitutes an asymptotically stable equilibrium neither on $\partial^{(m)} \mathscr{S}_{m+1}$ nor on $\partial^{(m+1)} \mathscr{S}_{m+1}$ corresponding to the assumption that these two boundary surfaces contain interior ESS's (i.e., unique attractors) of the associate m-strategy subgames; the latter condition $\bar{Q}'_{l'} = 0$ is necessary because $\bar{Q}' \in \partial \mathscr{S}_{m+1}$ must hold. Without loss in generality, let $\bar{Q}'_1 = 0$. For reasons of symmetry, $\bar{Q}'$ runs into a uniquely determined equilibrium point $\langle 0, \bar{Q}^0_2, \ldots, \bar{Q}^0_{m-1}, \bar{Q}^0_m, \bar{Q}^0_m \rangle$, $\bar{Q}^0_m \neq 0$, for $b \to 0$. Then $\bar{Q}^0 = \langle 0, \bar{Q}^0_2, \ldots, \bar{Q}^0_{m-1}, 2\bar{Q}^0_m, 0 \rangle$ is clearly an equilibrium of both the degenerate ($b = 0$) and the non-degenerate ($b \neq 0$) $(m+1)$-strategy game, too. Since $\bar{P}$ is an interior ESS of the unperturbed m-strategy game, it is the sole attractor of the flow on $\partial^{(m+1)} \mathscr{S}_{m+1}$. Hence $\bar{Q}^0$ is unstable, meaning that there exists a sequence $X^{(k)}$ ($k = 1, 2, \ldots$) in $\partial^{(m+1)} \mathscr{S}_{m+1}$ with non-vanishing E_1-components $X^{(k)}_1$ for which $\lim_{k \to \infty} X^{(k)} = \bar{Q}^0$ and $dX^{(k)}_1/dt > 0$. Then the sequence $Y^{(k)} = X^{(k)} + \bar{Q}' - \bar{Q}^0$ with $\lim_{k \to \infty} Y^{(k)} = \bar{Q}'$ satisfies $Y^{(k)} \in \mathscr{S}_{m+1}$ and $dY^{(k)}_1/dt > 0$ for all sufficiently large k. This contradicts the assumption that $\bar{Q}'$ is an attractor of the full $(m+1)$-strategy game. Hence $\bar{Q}$ is the sole attractor of the flow on $\mathscr{S}_{m+1}$.

Another way of proving the case $n = m + 1$ is to show that $\langle \bar{P}_1, \ldots, \bar{P}_m, 0 \rangle$ is a boundary point of the domain of attraction of $\bar{Q}$. Choose the Lyapounov function (LaSalle and Lefshetz 1961)

$$L(Y) = \prod_{j=1}^{m+1} Y_j^{\bar{Q}_j}$$

with

$$\frac{dL}{dt} = \bar{Q} \cdot BY - Y \cdot BY > 0$$

in a neighbourhood U of $\bar{Q}$, $\bar{Q} \neq Y$ (Schuster et al. 1981). If for $0 < z \ll 1$ $Z = \langle \bar{P}_1, \ldots, \bar{P}_{m-1}, \bar{P}_m - z, z \rangle$ is the population strategy in the moment when the mutation occurs, one finds that

$$\frac{dL(Z)}{dt} = \bar{P}_m \sum_{j=1}^m b_{m+1,j} \bar{P}_j = \bar{P}_m \bar{c}_m.$$

in linear approximation, where (V.28) has been used. Since $\bar{c}_{m'}$ is positive by assumption and $L(Z)$ is continuous in $Z, Z \in U$ for sufficiently small perturbations. A fortiori Z lies within the domain of attraction of $\bar{Q}$.

Let $n = k + m +$ now with $k \geq 1$, and assume that the proposition is valid for $n = k + m$. Analogously to the proof for the case $n = m + 1$, one shows that all solution trajectories in the interior of $\mathscr{S}_{k+m+1}$ flow to $\partial \mathscr{S}_{k+m+1}$. Hence the proof for $n = k + m + 1$ reduces to the case $n = k + m$, QED.

The Lemma states that the population game (IV.4) with the mixed ESS $\bar{P}$ will be transformed essentially according to Theorem 1 and its Corollary (Sect. IV.2.2) if $\bar{P}$ and C become perturbed by favourable mutants of one of the component strategies of $\bar{P}$. After such a perturbation, the final population state is again a mixed ESS with m non-vanishing components. The following theorem generalises this result to more than one mutating component of $\bar{P}$.

Theorem 2. Let $\bar{P}$ be an ESS of (IV.4) with $\bar{P}_j > 0$ for every j, $1 \leq j \leq m$. Assume $n - m$ additional strategy components that are mutants of the strategy components of $\bar{P}$ ($n \geq m + 1$). Let each of the mutants have a positive initial growth rate corresponding to (V.27) and (V.30). Then there exists a uniquely determined pay-off perturbation (m, m)-matrix c so that the perturbed system evolves into the ESS $\bar{P} + p$, whereby p and c correspond as in (IV.7).

Proof. The Lemma applies separately to the respective mutants of each particular strategy component of $\bar{P}$. So for every component $\bar{P}_i$ of $\bar{P}$, $1 \leq i \leq m$, there is at most one asymptotically stable mutant replacing its ancestral component strategy. The pay-offs associated with the total of these mutants contribute in a row-by-row fashion to the perturbation matrix c. The rest follows immediately from Theorem 1 and its Corollary.

Taylor (1979), Schuster et al. (1981) and Bomze et al. (1983) have treated the analogue of (IV.4) for asymmetric games in the sense of Maynard Smith and Parker (1976). Using the representation of asymmetric games given in Schuster et al. (1981), one straightforwardly modifies the above results so as to extend them to asymmetric dynamic games.

Remark 3. It seems unlikely that a result similar to Theorem 2 holds for non-ESS attractors even if the latter are structurally stable as in Zeeman (1980). For non-ESS attractors, (IV.7) does not generally determine the perturbation p uniquely. In such cases it is not clear which asymptotic state the system will attain when novel component strategies arise from favourable mutations.

Thus, when applied to dynamic evolutionary games, the condition (V.30) of structural instability has the following consequences. Up to small perturbations in pay-off and average population phenotypes, an ESS resists invasion by mutant component strategies of arbitrary kind and number provided selection is weak (Theorem 2). If the unperturbed ESS is based on m ($m \geq 1$) distinct behavioural phenotypes ("strategy components"), mutational perturbations of the average population phenotype are again m-strategy ESS's (Theorem 1 and Corollary, Sect. IV.2.2). Thus ESS's remain stable even in randomly changing genetic and ecological environments as long as the resulting variations in fitness remain sufficiently small. On the other hand, stable equilibrium strategies that are not ESS's may become destabilised by infinitesimal parameter disturbances due to genetic or environmental fluctuations affecting fitness (Remarks 1–2, Sect. IV.2.2; Remark 3).

We discuss a few biological implications of these results. As an example, consider a population whose genetic system and characteristic environment admit an ESS of intraspecific competition. According to (IV.4), the population evolves towards the ESS in dynamical time. Let the characteristics of the environment now change continuously in geological time, and let the gene pool be susceptible to random mutations modifying fitness. Thus novel strategies may arise against which the ESS has never been tested before. In view of the "practically universal pleiotropy" (Wright 1980) of genetic effects on phenotype which tends to average out most allelic difference, even favourable mutants will then modify only slightly the fitness pay-offs already present in the population game. Consequently, the ESS will be maintained during finite periods of secular evolutionary change. Alternatively, equilibria are unlikely to persist unless they are evolutionarily stable in the sense of game theory.

Maynard Smith (1982) and others have examined cases of diploid polymorphic populations in which genetic constraints do not admit an ESS. The structural-stability analysis suggests, however, that genetic evolution may show a general tendency to modify the constraints so as to produce ESS's only. A similar conclusion has been drawn by Thomas (1985a, b) from an explicitly genetic analysis of evolutionary stability. According to this analysis, a population equilibrium can always be destabilised by genetic mutations unless the population has arrived at a phenotypic ESS.

"In this sense, a phenotypic ESS is ultimately stable or 'uninvadable', and it takes new strategic components or alteration of payoffs to destabilize it."

(Thomas 1985a) The present theorems assert, however, that in case of weak selection not even novel strategy components or variations in pay-off will destabilise some given ESS.

Doubts of the theoretical significance of the ESS concept arose when Taylor and Jonker (1978) succeeded in showing that there are asymptotically stable equilibria which are not ESS's. More examples of this kind have been added by Zeeman (1980, 1981), Schuster et al. (1981) and others. Taylor and Jonker found the result "counterintuitive". Zeeman (1981) concluded that the concept of the asymptotic stability of population equilibria is a more general notion than evolutionary stability for characterising the ability to resist mutation. However, the present results suggest that asymptotically stable equilibria that are not ESS's may be structurally unstable, that is, transient from an evolutionary point of view.

1.8 Informal Summary and Conclusion

It has been argued that the evolutionary, dynamical approach to differential behaviour in the social animals can be extended so as to specify ways in which mutations that occur at random may affect the hereditary basis of behavioural adaptations. The argument has been based on strongly simplifying assumptions about the mechanisms of organic evolution which, however, are standard in mathematical population theory. The present approach may therefore be useful at least in clarifying the logic of sociobiological reasoning and evolutionary explanations of animal social behaviour that have previously been attempted on similar premises.

Criticisms of sociobiology concentrate largely on the fact that the genetic structures governing social behaviour are almost entirely unknown and that it is very unlikely that there exist one or a few alleles which express themselves in phenotypes of high individual flexibility as are characteristic of behavioural traits (e.g., Lewontin 1979, 1983; Montagu 1980). Contrary to that assumed in many sociobiological theories, Wright (1980) re-emphasised the notion that natural selection is primarily non-genic. In reply, Dawkins (1982, Chap. 2) argued that

"a gene for a trait X" simply means "there is genetic variation in the population for X",

according to current terminology in population genetics.

"When we use single-gene language in our adaptive hypotheses, we do not intend to make a point about single-gene models. We are usually making a point about *gene* models as against non-gene models."

(Dawkins 1982, p. 21) However, Lewontin (1982) remarked that evolutionary explanations in terms of genetic adaptations are hard to test in quite a fundamental sense because we could never know what ranges of variation were available to species during the course of their evolution.

The present investigation suggests that the interpretation of the basic sociobiological concepts need not be restricted to the selection of specific alleles for each social pattern. Nor is it necessary to assume similar genetic structures in different species from which analogous social responses are selected as independent adaptations to similar environmental features. However simplified and restrictive the sociobiological selection criteria and their underlying genetic assumptions may be, they do provide useful information for a theoretical understanding of *analogous* social patterns in evolutionary terms. The rationale of this situation lies in the broader significance of the selection criterion (V.30) covering and extending the results familiar from inclusive-fitness theory and the theory of ESS's. In fact, (V.30) specifies conditions under which the genomes of quite different species may independently produce the necessary mutational variation for one and the same adaptive trait.

2 Structural Instability in Population Dynamics

In recent years certain views of organic evolution have returned to prominence which reflect aspects of 19th-century evolutionary saltationism. Their late revival is sometimes interpreted as an attack on the fundamentals of neo-Darwinian theory since the synthesis of Mendelism and Darwinism decades ago has strongly reinforced the gradualist view of organic evolution among evolutionary biologists. However, the basic concepts of recent saltationism, such as "quantum evolution" (Simpson 1944), "genetic revolution" (Mayr 1954), "genetic transiliance" (Templeton 1980a, b), and "punctuational change" (Eldredge and Gould 1972; Gould and Eldredge 1977), have been introduced and discussed largely from the perspectives of modern synthetic theory. So the punctuational interpretation of natural history and the fossil record seems to be somewhat inconclusive as far as it relates to the question of whether instant speciation and related phenomena demand evolutionary mechanisms alternative to Darwinian natural selection (Gould and Eldredge 1977; Gould 1980, 1982). On the one hand, the punctuational hypotheses prove difficult to maintain in the light of microevolutionary accounts of continuously shifting gene frequencies in populations, which are often accepted as paradigms of speciation processes and thus lend much support to theories of macroevolutionary gradualism (Carson 1975). In fact, the paradigmatic role of population-genetic gradualism in macroevolutionary theory appears to be justified a posteriori by achievements such as Fisher's Fundamental Theorem of natural selection which explicates important features of Darwinian evolution within the gradualist conceptual framework (Ewens 1979, Chap. 1). Hence, phyletic gradualism has a serious, though debatable, theoretical base beyond those a priori assumptions and metaphysical prejudices of Western philosophy and culture that have been criticised by Gould and Eldredge (1977).

On the other hand, theoretical difficulties in reconciling the empirical evidence of evolutionary discontinuities with the approved mathematical framework of population genetics and ecological interactions arise from the fact that this framework is essentially restricted to relaxation processes in disequilibrium gene pools and ecosystems. The restriction is inherent in both the deterministic and the stochastic approach to population dynamics (Ewens 1979; Goel et al. 1971). To be sure, dynamic relaxation is not always a smooth directional process, and all sorts of instabilities–cycling, chaotic behaviour and irregular stochastic effects (random drift)—may occur in systems of interacting genotypes and phenotypes (Akin 1979; Ewens 1979; Goel et al. 1971; Olsen and Degn 1985). But since by definition selection due to differential reproduction and resource competition merely constitutes processes in dynamic time scales, with *secular variations* in the biomolecular and environmental constraints on these processes being systematically neglected, the relevance of the modes and patterns of dynamic relaxation to macroevolution may indeed seem questionable.

In the following sections it is argued, however, that macroevolutionary discontinuities are not only consistent with the basic assumptions of gradualism but even predicted – under certain circumstances – by the continuous-time models of population dynamics if the analysis is extended to instances of secular instability. Among the various approaches to related problems that have previously been attempted, we refer to Dodson (1976) who has presented a rigorous, though qualitative, mathematical treatment of morphological discontinuities arising from steady environmental changes in geological time, with the use of catastrophe-theory techniques. Under the assumption of moderate intraspecific evolution, Dodson could show that the "shifting peak" process familiar to population biologists implies long periods of morphological homeostasis interrupted by instances of rapid evolutionary change, whereby the latter results from stability thresholds in the "phenotype manifold", which, roughly speaking, is defined as the set of fitness extrema and neutrally stable points in adaptive landscapes. Dodson then demonstrated that his account of the shifting-peak process may explain the morphological gaps in the fossil record which the hypothesis of punctuated equilibria emphasises.

Various limits of the shifting-peak approach to evolutionary saltationism leave it worthwhile to reconsider the punctuated-equilibria model from the standpoint of population dynamics. It would seem obvious that the rate at which, and extent to which, the mean population phenotype responds to environmental changes depends, not only on the mere presence of sufficient morphological and genetic variability, but also on the structure and intensity of intraspecific competition and selection. Hence, whether peak shifts proceed gradually or arise instantaneously in geological time may prove highly contingent upon each population-dynamical system. On the other hand, evolutionary saltationism in the sense of Eldredge and Gould makes much stronger claims than the simple assertion of instantaneous morphological transformation, which have received little mathematical treatment yet. The following investigation suggests that if interactions within and between populations are sufficiently complex (e.g., with effects of density-dependent selection and sympatric evolution included), periods of relative morphological stasis (punctuated equilibria) alternating with finite evolutionary steps become prominent features of macroevolutionary change.

2.1 Population Interactions in Secular Time Scales

In order to describe intraspecific population interactions, we classify the members of a species into m distinct categories (genotypes, phenotypes, subpopulations, etc.) of size $X_i \geq 0$, $1 \leq i \leq m$, whose reproduction rates are given by polynomials $G_i(X_1, \ldots, X_m)$ of degree $n \geq 1$, and obey the differential equations

$$\frac{dX_i}{dt} = X_i G_i(X_1, \ldots, X_m), \quad 1 \leq i \leq m. \tag{V.41}$$

It is assumed that the coefficients entering G_i are constant with regard to dynamical time scales, and spatially homogeneous. For $n = 1$, Eqs. (V.41) are the Lotka-Volterra equations of m interacting populations, with constraints imposed on G_i corresponding to predation, competition, symbiosis or ecological saturation (e.g., Roughgarden 1979). For $n = 2$,

$$G_i(X_1, \ldots, X_m) = A_i + \sum_{j=1}^{m} B_{ij} X_j + \sum_{j,l=1}^{m} C_{ijl} X_j X_l \tag{V.42}$$

covers strongly density-dependent population interactions, while the cases $A_i = 0$, $C_{ijl} = -\delta_{ii} B_{jl}$ and $\sum_{j=1}^{m} X_j = 1$ lead to the basic equations of population genetics of the classical Fisher-Haldane-Wright type, and to evolutionary game dynamics (see previous sections). However, predator-prey interactions and, except for one or the other occasional remark, population genetics will not be considered in this section.

Note that polynomials of higher degrees tend to yield structurally more complex systems in the strict sense of Part One. For example, the case $n = 2$ can be viewed as the coupling ($C_{ijl} \neq 0$) of previously decoupled systems ($C_{ijl} = 0$) with $n = 1$.

General interest in mathematical population dynamics concentrates on the topological structure of the phase portraits of the solution trajectories $X_i(t)$ – in particular, on the domains of attraction of stable equilibria of (V.41), stable limit cycles, and the like (Roughgarden 1979, Part Five; Hofbauer 1981). However, since the flow $X_i(t)$ does not describe processes other than ecological relaxation, a structural-instability approach to (V.41) is attempted here in order to analyse macroevolutionary modes and patterns. Structural-stability analyses of solution trajectories $X_i(t)$ have previously been carried out by Zeeman (1980, 1981), Thomas and Pohley (1932) and Brown (1983) in evolutionary game dynamics, and by Akin (1979) in population genetics, and in the context of numerous problems in theoretical ecology (see various contributions to Gurel and Rössler 1979). We start from the obvious assumption that the coefficients ("control parameters") A_i, B_{ij}, C_{ijl} and possible contributions to G_i of higher degrees arise from epigenetic interactions between the genome and the characteristic environment of a population, thus specifying reproductive differences between species-specific phenotypes. It is further assumed that long-term random genomic substitutions as well as environmental variations may lead to gradual changes in the control parameters. The time average of (V.41) over sufficiently long ecological periods is taken in some well-defined sense, and thus the "phenotype profile"

$$f_i := \frac{\bar{X}_i}{\sum_{j=1}^{m} \bar{X}_j}, \quad 1 \leq i \leq m,$$

and mean phenotype of populations or species are obtained, where the bars in $\bar{X}_i$ indicate the time average. Now, the long-term evolution of polymorphic populations can be described by variations in f_i, with the control parameters $A, B, C, \ldots$ varying on secular time scales.

It should be noted that the averaging procedure suggested here is not only suitable for eliminating transient states from dynamical systems (V.41) which are of no long term evolutionary significance, but also helpful in interpreting (V.41) in view of Smale's (1976) and Akin's (1979) results. These authors have shown that under quite realistic conditions certain systems of the type (V.41), frequently considered in competitive-selection theory and populations genetics, may prove compatible with arbitrary dynamical behaviour provided m is large. As will become evident below, however, the time averages even of dynamically unstable trajectories may still be well-defined features of the system (V.41) in many physically relevant cases.

In order to calculate $\bar{X}_i$ from the time average of (V.41), we introduce a nonsingularity condition for $X_i(t)$. We first consider trajectories that remain bounded away from faces $X_j(t) \equiv 0$ for some $j \leq m$, which means that for some $\varepsilon > 0$ we have $X_i(t) > \varepsilon$ for all $i \leq m$ and sufficiently large t. The limiting case $\varepsilon \to 0$ is then treated separately. We also assume finite carrying capacities $K > 0$ for natural environments so that $G_i < 0$ for $X_i > K, i \leq m$, which prevents $X_i(t)$ from growing indefinitely. Under these conditions (V.41) gives, for appropriate initial values $X_i(0) \neq 0$,

$$\lim_{t \to \infty} t^{-1}[\ln X_i(t) - \ln X_i(0)] = = \lim_{t \to \infty} t^{-1} \int_0^t P_i(X_1, \ldots, X_m) dT$$

$$= \overline{P_i(X_1, \ldots, X_m)} = 0. \tag{V.43}$$

The case $n = 1$ with

$$\sum_{j=1}^{m} B_{ij} \bar{X}_j = -A_i \tag{V.44}$$

has been treated by Goel et al. (1971). The solutions $\bar{X}_i$ are identical to the equilibria of (V.41). They exist if det $\| B_{ij} \| = D \neq 0$. Defining $D_i = $ det $\| B'_{ij} \|$ with A_i replacing the ith column in $\| B_{ij} \|$ so that the matrix $\| B'_{ij} \|$ results, one has the unique solution

$$\bar{X}_i = D_i/D, \quad 1 \leq i \leq m, \tag{V.45}$$

according to Kramer's rule. Note that for $D = 0$ solutions exist only if they are degenerate in a trivial sense (all $D_i = 0$). Systems (V.44) with $\bar{X}_j < 0$ for at least one index j are not considered in order to exclude unphysical cases (choose $\bar{X}_j = 0$ and reduce the rank of $\| B_{ij} \|$ appropriately). The phenotype frequency distribution of the population now reads

$$f_i = \frac{D_i}{\sum_{j=1}^{m} D_j} > 0. \quad 1 \leq i \leq m. \tag{V.46}$$

Since D and D_i are smooth functions of A_i and B_{ij}, the frequencies f_i vary smoothly in those parameter regimes $R \subset \mathbb{R}^3$ of A_i, B_{ij} which admit physical, non-degenerate solutions of (V.44). This holds even if the condition $D_i \neq 0$ is eventually relaxed and parameter variations possibly leading to $D_i \to 0$, $\bar{X}_i \to 0$ are included. Parameter values for which $D_i = 0$ indicate bifurcation points of the system (V.41), where profiles $f_1, \ldots, f_m$ corresponding to different regimes $R \subset \mathbb{R}^3$ match in a continuous way (see bifurcations in Fig. V.2). One thus has the result that the averages of X_i over ecological periods are structurally stable functions of the control parameters A_i, B_{ij} in the sense that when A_i, B_{ij} vary gradually in geological time, so will the phenotype profiles of populations and species. Using the familiar case with $m = 2$, the significance of this result for evolutionary fitness peak shifts is discussed in the next section.

For $n = 2$, we first treat the important special case $A_i = 0$, $1 \leq i \leq m$, with $C_{ijk} = -\delta_{ii} B_{jk}$ and $\sum_{j=1}^{m} X_j(t) = 1$ (game dynamics). Hofbauer (1981) has shown that the system (V.41) is then essentially equivalent to a system of Lotka-Volterra equations with $n = 1$, so that the above results extend to the game-dynamic case as well (cf. Zeeman 1980, p. 481). Suppose now $A_i \neq 0$ for at least one index i, with $\sum_{j=1}^{m} X_j(t) = 1$ not being necessarily required. The non-singularity condition assumed in (V.43) then leads to

$$\sum_{j=1}^{m} B_{ij} \overline{X_j} + \sum_{j,l=1}^{m} C_{ijl} \overline{X_j X_l} = -A_i, \quad 1 \leq i \leq m, \tag{V.47}$$

Clearly (V.47) still applies when $\lim_{t \to \infty} X_j(t) = 0$ for some index j. The system (V.47) has m equations for the $m(m+3)/2$ unknown quantities $\overline{X_i}$, $\overline{X_i X_j}$ and is thus underdetermined. For a given set of parameter values, one must generally expect a multiplicity of different sets of trajectories $X_i(t)$ which fulfil the average equations. The average phenotypes of polymorphic populations become irremovably codetermined by the initial conditions of the averaging procedure and may change spontaneously in a discontinuous fashion under infinitesimal perturbations of the trajectories.

We illustrate the case $n = 2$ by an example in which the right-hand side of (V.41) takes the form of a gradient

$$\frac{\partial V}{\partial X_i} = X_i P_i = X_i \left(A_i + B_i X_i + \sum_{j=1}^{m} C_{ij} X_j^2 \right) \tag{V.48a}$$

with potential

$$V = \frac{1}{2} \sum_{j=1}^{m} A_j X_j^2 + \frac{1}{3} \sum_{j=1}^{m} B_j X_j^3 + \frac{1}{4} \sum_{j,l=1}^{m} C_{jl} X_j^2 X_l^2 \tag{V.48b}$$

and $C_{jl} = C_{lj}$. Up to an additive constant in V, the special form (V.48) of V and G_i is uniquely determined by the integrability conditions $\partial V/\partial X_i = X_i G_i$ and H_{ij}

$= \partial^2 V/\partial X_i \partial X_j = \partial^2 V/\partial X_j \partial X_i$ (symmetry of the Hessian matrix $\|H_{ij}\|$ of V). A low-dimensional close analogue of (V.48) ($m = 2$), well known in population dynamics, will be discussed below. Because of the symmetry of the Hessian matrix (purely real eigenvalues), Hopf bifurcations and cycling cannot occur. Since one also has $dV/dt \geq 0$, the limit $\lim_{t \to \infty} X_i(t) = \bar{X}_i$ (equilibrium, stable or unstable) always exists. L'Hôpital's rules then stipulate the relations

$$\overline{X_i(t)} = \lim_{t \to \infty} t^{-1} \int_0^t X_i(T) dT = \lim_{t \to \infty} X_i(t) = \bar{X}_i$$

and, similarly, $\overline{X_i(t)^2} = \bar{X}_i^2$. Altogether, this reduces the average equations (V.47) to the stationarity conditions for (V.41) which adopt the form

$$B_i \bar{X}_i + \sum_{j=1}^m C_{ij} \bar{X}_j = -A_i, \quad 1 \leq i \leq m. \tag{V.49}$$

This result also holds in the more general case in which a Lyapounov function exists for (V.41).

The Hessian determinant reads

$$\det \|\bar{H}_{ij}\| = \left[\prod_{j=1}^m \bar{X}_j\right] \det \|B_i \delta_{ij} + 2C_{ij} \bar{X}_j\|$$

$$= \Delta \prod_{j=1}^m \bar{X}_j. \tag{V.50}$$

The determinant Δ is a polynomial of degree m in $\bar{X}_i$. Let $\bar{X}_i$ change, for instance, according to secular variations in A_i. Then the Hessian determinant will generally have domains in which it is non-singular, and subsets in $(\mathbb{R}_0^+)^m$ in which it is zero. The set of parameter values for which the Hessian determinant vanishes is known as the "bifurcation set" (Poston and Stewart 1978). It cuts the control space into separate regimes with distinct stability properties and multiplicities of the equilibria $\bar{X}_i$. The resulting modes of evolution in secular time scales are well known from dynamical systems theory. Slow continuous variations in the control parameters forcing the system to pass through the bifurcation set at points with $\Delta = 0$ (degenerate critical points) may induce structural instabilities of the catastrophe type in the phenotype profiles

$$f_i = \frac{\bar{X}_i}{\sum_{j=1}^m \bar{X}_j}$$

of evolving species. The conclusion is that if in polymorphic systems competitive and cooperative interactions are sufficiently complex, gradual and discontinuous macro-evolutionary transitions in average morphological structure will generally alternate.

We add a few qualifying remarks. Evidently, the latter conclusion holds a fortiori for models with $n \geq 3$. When no potential function exists for (V.41), the structural stability analysis may become complicated, for cycles, "strange attractors", and chaotic behaviour must be expected to arise. This, however, may be expected to support rather than invalidate the general conclusion on the occurrence of evolutionary discontinuities that has been drawn in the potential-function case.

Similarly, when G_i is assumed to constitute a type of more complicated functions than the polynomials considered here (e.g., with spatial inhomogeneities, stochastic dynamical effects, etc. admitted), secular instabilities of population structure will be no less pronounced in such systems. Of the many possibilities one may think of, we only mention the analysis by Oster et al. (1976) of density-dependent discrete-time systems in which shifts in genotype distribution tend to generate bifurcations and periodic and chaotic behaviour in the average morphological structure of populations and ecosystems.

2.2 Low-Dimensional Examples

We discuss the results obtained thus far in the light of a few competition equations familiar in theoretical ecology. Although the equations are mathematically simple, they illustrate the modes of both gradual and punctuational evolution in a straightforward fashion.

(A) $n = 1, m = 2$

Consider a species or population S_1 of size X_1 which in a steady and homogeneous environment obeys an equation of logistic growth

$$\frac{dX_1}{dt} = k_1(K_1 - X_1)X_1, \tag{V.51}$$

where the coefficients k_1 and K_1 are related to the birth and death rates, and to the available amount of food and other resources. Let a mutant subpopulation S_2 now appear in S_1, and let S_2 be characterised by population size X_2 and saturation level K_2. The organisms of type S_2 compete with the non-mutants for a share β in the resources present in the ecosystem, where $0 \leq \beta \leq 1$. If $\beta = 1$ this implies identical ecological niches for S_1 and S_2, whereas $0 \leq \beta < 1$ indicates a partial or total differentiation in the respective resources. Then the interaction of S_1 and S_2 is described by

$$\frac{dX_1}{dt} = k_1 X_1 (K_1 - X_1 - \beta X_2), \tag{V.52a}$$

$$\frac{dX_2}{dt} = k_2 X_2 (K_2 - X_2 - \beta X_1). \tag{V.52b}$$

The ecosystem (V.52) has frequently been analysed in theoretical population dynamics (e.g., Roughgarden 1979). For the sake of completeness of the following discussion, a few results must be summarised briefly here. Shortly before the mutant S_2 appears, the steady solution $\bar{X}_1 = K_1, \bar{X}_2 = 0$ obtains, and (V.52) reduces to (V.51). The system is unstable against invasion by S_2 if

$$K_2 > \beta K_1. \tag{V.53}$$

If one also has $\beta = 1$, then S_2 completely replaces S_1, and the final population number is exclusively given by $\bar{X}_2 = K_2 > K_1$. In case of $\beta = 0$, S_1 coexists with S_2 and the total population number is increased from $\bar{X}_1 = K_1$ to $\bar{X}_1 + \bar{X}_2 = K_1 + K_2$. In the intermediate range of overlapping ecological niches ($0 < \beta < 1$), two cases must be distinguished. If

$$\beta K_2 > K_1 \tag{V.54}$$

in addition to (V.53), again S_1 becomes extinct, but the final population size $\bar{X}_2 = K_2$ is still larger than the initial one. On the other hand, in case

$$\beta K_2 < K_1, \tag{V.55}$$

both subspecies coexist:

$$\bar{X}_1 = \frac{K_1 - \beta K_2}{1 - \beta^2},$$
$$\bar{X}_2 = \frac{K_2 - \beta K_1}{1 - \beta^2}, \tag{V.56}$$

so that

$$\bar{X}_1 + \bar{X}_2 = \frac{K_1 + K_2}{1 + \beta} > K_1. \tag{V.57}$$

The ecosystem (V.52) explicates basic results of Darwinian theory. Evolution proceeds so as to lead to an increasing exploitation of resources and ecological niches. Successive mutations will be rejected if they do not increase the overall degree of adaptation (expressed as the equilibrium size) of the populations involved. Niche variation ($\beta < 1$) is favoured when new resources become accessible which are rich enough to sustain polymorphism. Otherwise only one of the competing subspecies will be selected for survival (competitive exclusion). A simple example of the long-term evolution of S_1 and S_2 may be given by genetic or environmental changes in secular time scales favouring S_2 in the sense that K_2 slowly increases with time, whereas K_1 and $\beta < 1$ are kept constant. The population numbers $\bar{X}_1, \bar{X}_2$, and $\bar{X}_{tot} = \bar{X}_1 + \bar{X}_2$ are plotted in Fig. V.2 as functions of K_2 according to (V.52) to (V.57). Figure V.2 shows

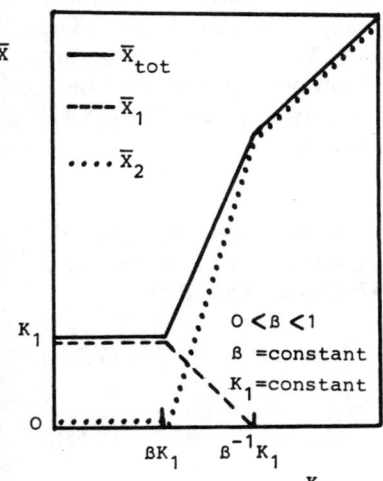

Fig. V.2. Population numbers $\bar{X}_1, \bar{X}_2$, and $\bar{X}_{tot} = \bar{X}_1 + \bar{X}_2$ as continuous functions of the saturation level K_2 according to (V.52). For secularly varying K_2 the evolution of the polymorphic system is purely gradual (after Geiger 1983)

that the dependent variables are continuous functions of the saturation level K_2. Corresponding to (V.44) and (V.45), this type of polymorphism evolves gradually as long as the control parameter K_2 changes continuously.

The system (V.52) is useful in delineating the significance of evolutionary discontinuities occurring in fitness peak-shift approaches. Dodson (1976) makes the obvious assumption that in general there will be more than one phenotype (i.e., subspecies S_i) maximising the overall population fitness locally relative to some given environment. Then the question arises as to which peak the fitness assumes, that is, which phenotype will be selected. In a changing environment, a uniform population in stable equilibrium with the environment can evolve towards a better-adapted phenotype only when the saddle between the old and the new peak becomes elevated and approximates the old peak in height. This increases the phenotypic variability in the population, with rapid motion towards the new peak eventually setting in. Dodson showed that this pattern of directional selection corresponds to structural instabilities in the "phenotype manifold" of an evolving species.

When stated in terms of the competition equations (V.52), an adaptive peak at phenotype S_1, say, with ecological saturation level K_1, implies a set of suboptimal phenotypes S_2 varying continuously around S_1, with parameters K_2 and β, $0 < \beta \leq 1$, so that

$$K_1 > \beta^{-1} K_2. \tag{V.53'}$$

The system (V.52) admits a Lyapounov function

$$L = K_1 X_1 + K_2 X_2 - \tfrac{1}{2}(X_1^2 + X_2^2) - \beta X_1 X_2 + const. > 0$$

with $dL/dt > 0$ for all $X_1 \neq \bar{X}_1 = K_1$, $X_2 \neq \bar{X}_2 = 0$. Evidently, L is the desired fitness function, having a maximum at S_1. Annihilation of the peak at S_1 consists in a reversal of the inequality sigh in (V.53'). This gives rise to a continuum of phenotypes S_2 intermediate between S_1 and S_1',

$$K_1 \leq \beta^{-1} K_2 \leq \beta'^{-1} K_1', \tag{V.53''}$$

where S_1' is some other local optimum phenotype distinct from S_1, with the peak at K_1' exceeding K_1 sufficiently in height. The fitness maximum then starts to move away from S_1 towards S_1', but the peak shift proceeds gradually as shown in Fig. V.2. The environment may indeed optimise distinct phenotypes $S_1, S_1', \ldots$ simultaneously, but the dynamics of the compound polymorphism will always maximise the fitness uniquely. This is a simple consequence of the fact that polymorphic systems (V.41) with $n = 1$ cannot have non-trivial simultaneous equilibria.

(B) $n = 2$, $m = 2$

The situation changes when more complex modes of competition are made explicit. The transformations $\beta \to \beta_1 X_2$ in (V.52a) and $\beta \to \beta_2 X_1$ in (V.52b) are carried out so that the resulting equations (Hutchinson 1947; Cunningham 1955)

$$\frac{dX_1}{dt} = k_1 X_1 (K_1 - X_1 - \beta_1 X_2^2), \tag{V.58a}$$

$$\frac{dX_2}{dt} = k_2 X_2 (K_2 - X_2 - \beta_2 X_1^2) \tag{V.58b}$$

describe two distinct subpopulations competing for limited resources in (partly) overlapping ecological niches, while the mixed interaction terms of the third degree indicate density dependence on the extent to which the niches overlap ($\beta \to \beta X$). The two phenotypes not only eat up the available resources in proportion to their respective numbers, but also invade each others' ecological niches more as they grow larger. This mode of interaction may obtain, for instance, when competition involves sympatry and habitat selection as a density-dependent process eventually leading to competitive exclusion (Partridge 1978).

We first draw the contrast with Eqs. (V.52). The system (V.58) gives

$$\bar{X}_1 = K_1 - \beta_1 \bar{X}_2^2 \tag{V.59a}$$

$$\bar{X}_2^4 - a\bar{X}_2^2 + b\bar{X}_2 + c = d, \tag{V.59b}$$

where the coefficients a, b, c, d (all positive) are easily obtained from (V.58) and (V.59a). We only note that d is proportional to K_2 and that for $K_1 \gg 1, \beta_1 < 1, \beta_2 < 1$ (which may be the most realistic cases) we have $c \gg a$. Hence (V.59b) cannot have positive roots. Similarly to Fig. V.2, $\bar{X}_1$ and $\bar{X}_2$ have been drawn schematically in Fig. V.3 as functions of the population parameter K_2, where

$$(\tfrac{4}{3} K_1)^3 \geq \beta_1^{-1} \beta_2^{-2} \tag{V.60}$$

has been assumed. By discussing the extrema of (V.59b) it can be easily verified that the loops in Fig. V.3a,b vanish if (V.60) is violated. Linear stability analysis of the

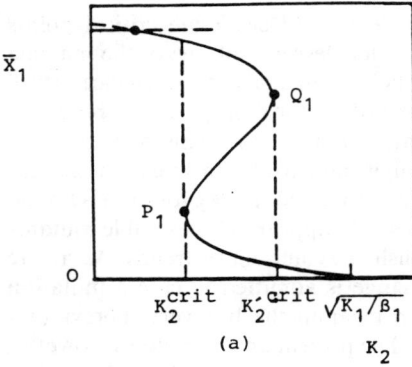

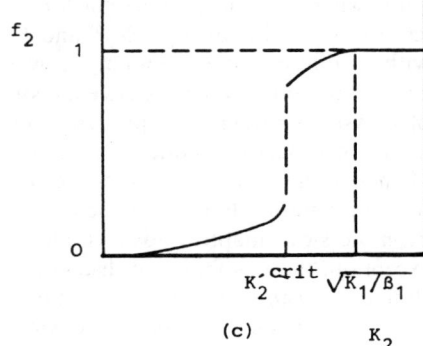

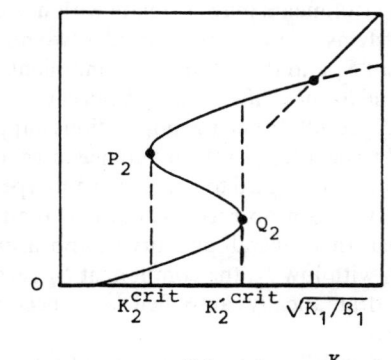

Fig. V.3a–c. Population numbers $\bar{X}_1$ (a) and $\bar{X}_2$ (b), and phenotype frequency f_2 (c) as functions of K_2 according to (V.58). The curves in a and b have two "catastrophe" points P and Q, respectively, where evolutionary saltations can occur. In addition, $\bar{X}_2$ has a bifurcation point for $K_2 = (K_1 \beta_1^{-1})^{1/2}$ when $\bar{X}_1$ vanishes for increasing K_2. Similarly, for increasing K_2 the phenotype frequency f_2 undergoes a marked jump at $K_2 = K_2'^{crit}$ on the lower stability branch of $\bar{X}_2$. Slight genetic variations leading to infinitesimal fluctuations in K_2 at $K_2'^{crit}$ will therefore have a large impact on the average phenotype of the population (after Geiger 1983)

stationary state (V.59) leads to the result that on the intermediate branches between P_1 and Q_1 (P_2 and Q_2) in Fig. V.3 the solution (V.59) is unstable, but it is stable elsewhere. For higher or lower K_2, respectively, these points indicate structurally unstable states. Variations of K_2 around its critical values subject one phenotype or subspecies to a "population flush", whereas the other is drastically reduced, For $K_2^2 = K_1 \beta_1^{-1}$, where S_1 becomes extinct for increasing K_2, Eq. (V.58b) has a bifurcation point at which (V.59b) is replaced by $\bar{X}_2 = K_2$. The growth rates of an infinitesimal perturbation of (V.59) can be easily derived from the linear perturbation analysis (Geiger 1983). On the stationarity curve between P_2 and Q_2 one of the growth rates adopts positive values, which implies instability. This is exactly true for the intermediate values of $\bar{X}_1$ and $\bar{X}_2$ between P and Q when the competition terms in (V.58) become large. If we again assume that S_2 is a mutant population of S_1, then the instability of the intermediate states means that for about equally large populations competition becomes too strong for pronounced polymorphism to be maintained. Again, the success of an evolutionary step depends on the relative degrees of adaptation between S_1 and the mutant S_2 as expressed by K_1 and K_2. However, the case $m = 2$ shows that the outcome of the contest may become subject to an almost complete inversion, forcing the population into instantaneous transitions between distant, secularly stable branches.

It is easy to see that (V.58) has the Lyapounov function

$$L = \beta_1 X_1^2 \left(\frac{K_1}{2} - \frac{X_1}{3} \right) + \beta_2 X_2^2 \left(\frac{K_2}{2} - \frac{X_2}{3} \right) - \frac{\beta_1 \beta_2 X_1^2 X_2^2}{2} + const.$$

with two simultaneous maxima for $K_2^{crit} < K_2 < K_2'^{crit}$. Degenerate critical points (cf. Fig. V.3) arise for $K_2 = K_2^{crit}$ and $K_2 = K_2'^{crit}$ (coalescence of one of the maxima with the intermediate minimum). However, in the present analysis coexistence as well as annihilation of fitness peaks expresses intrinsic dynamical properties of polymorphic systems rather than particular environmental situations. This may generate modes of unsteady evolution even in constant environments. According to the axioms of pheletic gradualism, in constant environments secular adaptive evolution is bound to change smoothly in the sense that spontaneously appearing favourable mutants promote slow adaptive advance along established evolutionary trends. As a rare exception, the possibility of discontinuous change is admitted in case a mutation should bring about a new adaptive type at a step, meaning the discovery of previously unknown ways of exploiting the environment. The present analysis shows, however, that under quite realistic conditions (sympatric evolution, density-dependent selection) secular genetic evolution in constant environments may correlate with marked phenotypic discontinuities in polymorphic systems. For instance, consider two mixed populations containing a common component S_1 and the respective components S_2 and $\hat{S}_2$ in certain ratios. The components S_2 and $\hat{S}_2$ may differ slightly from each other in their genetic makeup so that variations in K_2 result. When the populations happen to range slightly below and above $K_2 = K_2'^{crit}$, according to the difference in genetic structure, they nonetheless appear quite distinct with regard to average phenotype, as can be seen by the step in f_2 shown in Fig. V.3c. An observer would find the two populations genetically rather similar, whereas their overall phenotypic appearance would differ considerably. In the population with low f_2, the component S_2 carries some hidden genetic variation which is dramatically amplified and becomes dominant when K_2 is increased above $K_2'^{crit}$.

The complete structural stability analysis of (V.58) is more conveniently carried out when dimensionless variables $Y_i = X_i/K_i$ and parameters $\kappa_i = k_i K_i$, $\alpha_1 = \beta_1 K_2^2 K_1^{-1}$, $\alpha_2 = \beta_2 K_1^2 K_2^{-1}$ are introduced so that

$$\frac{dY_1}{dt} = \kappa_1 Y_1 (1 - Y_1 - \alpha_1 Y_2^2), \tag{V.61a}$$

$$\frac{dY_2}{dt} = \kappa_2 Y_2 (1 - Y_2 - \alpha_2 Y_1^2). \tag{V.61b}$$

The Lyapounov function being redefined accordingly, the vanishing of its Hessian determinant at equilibria

$$\bar{Y}_1 = 1 - \alpha_1 \bar{Y}_2 \tag{V.62a}$$

$$\bar{Y}_2 = 1 - \alpha_2 \bar{Y}_1 \tag{V.62b}$$

leads to

$$1 - 4\alpha_1 \alpha_2 \bar{Y}_1 \bar{Y}_2 = 0$$

and

$$\bar{Y}_1 \bar{Y}_2 = 4(1 - \bar{Y}_1)(1 - \bar{Y}_2).$$

Choosing the parameterisation

$$\frac{\bar{Y}_2}{1 - \bar{Y}_2} = 2\tau, \qquad \frac{\bar{Y}_1}{1 - \bar{Y}_1} = 2\tau^{-1},$$

one substitutes back into (V.62) and gets

$$\alpha_1 = \frac{(1 + 2\tau)^2}{4\tau(\tau + 2)}, \qquad \alpha_2 = \frac{(\tau + 2)^2}{4(1 + 2\tau)}$$

for the bifurcation set shown in Fig. V.4. The curves intersecting at $\alpha_1 = \alpha_2 = 3/4$ (see

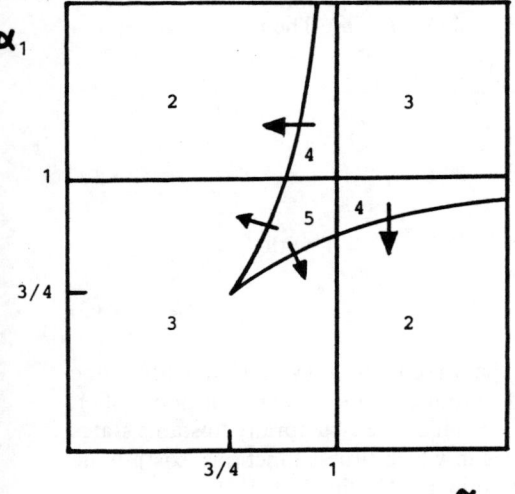

Fig. V.4. Bifurcation diagram of the system (V.61). The *numbers* give the multiplicities of equilibria $\bar{Y}_1, \bar{Y}_2$ (stable and unstable) arising in the various parameter regimes. The *arrows* indicate parameter changes involving structural instabilities of the average phenotype in polymorphic systems (after Geiger 1983)

(V.60)) closely resemble the bifurcation diagram of the canonical "cusp catastrophe" (Thom 1975; Poston and Stewart 1978). The numbers in the various parameter regimes give the total numbers of stable and unstable equilibria $(\bar{Y}_1, \bar{Y}_2)$ in these regimes. Crossing the curved lines in the indicated directions leads to discontinuities in the average population phenotype such as have been discussed in the context of Fig. V.3. Figure V.4 shows that even in simple cases such as (V.58) or (V.61), involving only two governing parameters, a surprising multiplicity of secularly unstable morphologies may arise in phyletic lines evolving across the α_1–α_2 plane.

2.3 The Adaptive Topography Reconstructed

For further analysis of the modes of long-term evolution, the concept of fitness surface, or adaptive topography, must be modified appropriately. The graph of the fitness function, usually accepted as its geometrical representation, is clearly unsuitable for macroevolutionary analyses since it is not invariant under secular changes. Moreover, it does not even exist when there is no Lyapounov function for (V.41) (cf. Akin 1979). Instead, we choose the surface of average values $\langle \bar{X}_1, \ldots, \bar{X}_m \rangle$, varying with the control parameters, as the most suitable representation of the adaptive surface. It coincides with the "catastrophe manifold" (Thom 1975; Poston and Stewart 1978) when (V.41) admits a Lyapounov function, and keeps working when not. Moreover, dynamic (transient) features of no long-term evolutionary significance thus become removed from the adaptive topography in a natural way. We finally replace the $\bar{X}_i$ by the phenotype frequencies f_i so that the morphological structure of mixed populations can be visualised immediately.

In fact, the well-known Implicit-Function Theorem shows that except for degenerate points the set of time averages of the solution trajectories is a locally smooth surface (subset of $(\mathbb{R}_0^+)^m$) defined by

$$\overline{P_i(X_1, \ldots, X_m)} = 0, \quad 1 \leq i \leq m, \tag{V.43'}$$

in the control space. Trivially, $\bar{P}_i$ in (V.43') is a smooth function of the $\bar{X}_i$ since the correlations $\overline{X_j X_l \ldots}$ entering $\bar{P}_i$ depend smoothly on $\bar{X}_1, \ldots, \bar{X}_m$. In order to see this, choose perturbation functions $x_h(t)$ with $\bar{x}_h \neq 0$ ($1 \leq h \leq m$). Then

$$\frac{\partial \overline{X_j X_l \ldots}}{\partial \bar{X}_h} = \lim_{\varepsilon \to 0} \frac{\overline{(X_j + \varepsilon \delta_{jh} x_j)(X_l + \varepsilon \delta_{lh} x_l) \ldots} - \overline{X_j X_l \ldots}}{\varepsilon \bar{X}_h}$$

$$= \frac{\delta_{jh} \overline{x_j X_l \ldots} + \delta_{lh} \overline{X_j x_l \ldots} + \ldots}{\bar{x}_h}$$

is a well-defined expression. For averages $\langle \bar{Y}_1, \ldots, \bar{Y}_m \rangle = \bar{Y}$ with

$$\det \left\| \frac{\partial \overline{P_i(X_1, \ldots, X_m)}}{\partial \bar{X}_j} \right\|_{\bar{X} = \bar{Y}} \neq 0 \tag{V.63}$$

the Implicit-Function Theorem then states that the equations (V.43') implicitly define $\bar{X}$ as a smooth function of the control parameters in a neighbourhood of $\bar{Y}$. Alternatively, violations of the condition (V.63) indicate structurally unstable states (including degenerate critical points in cases in which fitness functions exist). Note that (V.63) is the generalisation of the determinant Δ defined in (V.50).

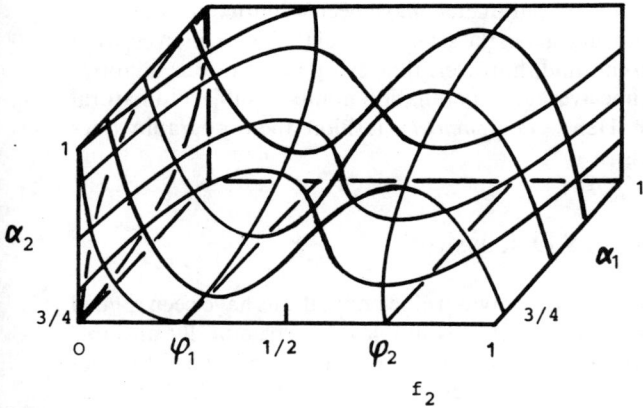

Fig. V.5. Adaptive topography of the ecosystem (V.61) according to (V.64), with bifurcation set given by the eye-shaped curve in the α_1–α_2 plane. Between the critical frequencies φ_1 and φ_2 there is a characteristic gap in the distribution of stable polymorphic states (redrawn from Geiger 1983)

For $n = 1$, the adaptive surface consists in a piecewise smooth hypersurface in $(\mathbb{R}_0^+)^m$, that is, pieces of smooth surfaces matching in a continuous fashion analogous to the low-dimensional example of Fig. V.2. For illustrative purposes, we choose the example (V.61). Rewriting $\bar{Y}_i = \bar{N} f_i$ ($i = 1, 2$) and eliminating $\bar{N}$ from (V.62) gives an implicit equation

$$(\alpha_1 f_2^2 - \alpha_2 f_1^2)^2 + (f_1 - f_2)(\alpha_1 f_2^3 - \alpha_2 f_1^3) = 0 \qquad (V.64)$$

for the adaptive surface shown in Fig. V.5. According to the symmetry of (V.64), there is a "forbidden" area of mean population phenotypes around $f_1 = f_2 = \frac{1}{2}$ that corresponds to the domain of unstable equilibria depicted in Fig. V.3. The bifurcation set is given by the closed eye-shaped curve in the α_1–α_2 plane.

In general, the adaptive topography must be expected to depend on more than two control parameters. It is well known from the theory of many-parameter catastrophe manifolds that the resulting geometries tend to be tricky to describe in detail. For the present purposes it may suffice, however, to infer qualitatively certain general properties of evolution on secular time scales from Fig. V.5. Firstly, the modes of gradual and unsteady evolution correspond respectively to the local and global topographical structure of adaptive surfaces. Secondly, it is similarity of the average population phenotype to the critical phenotypes φ_1 and φ_2, not morphological flexibility, that favours finite morphological steps. Morphological stasis prevails in populations sufficiently dissimilar to φ_1 and φ_2, or equivalently, in the structural stability regimes where finite parameter variations yield little change in mean phenotype. On the other hand, the probability of instantaneous steps strongly increases in the vicinity of secularly unstable polymorphic states, where even infinitesimal parameter perturbations due to genetic or environmental fluctuations may have major effects on the mean population phenotype. Finally, in higher-dimensional cases with $m > 2$ the directions of finite steps need not coincide with trends of steady morphological evolution. The adaptive topography may then contain critical points of multiple degeneracy with regard to $f_1, f_2, \ldots$ (Poston and Stewart 1978, Chap. 7).

A final word should be said about the present use of the term "adaptive topography". The preceding analysis has been based on the familiar assumption that natural selection and adaptation can be adequately described as dynamic processes, whereas secular changes in the control parameters have been admitted to constitute modes of non-adaptive evolution, at least in the strict sense of the Darwinian concept of adaptation. One must keep in mind, however, that every point in our adaptive topography corresponds to a time average of population numbers subject to natural selection and, thus, adaptation. Hence, in a *pointwise* fashion, the topography does characterise adaptation processes.

2.4 Informal Summary and Discussion

Intraspecific competition and selection in polymorphic populations have been argued to be inherently non-linear processes which possibly lead to structurally unstable states in species and ecosystems. Discontinuous genetic and morphological changes may thus occur, which can be interpreted as evolutionary "catastrophes", that is, instantaneous transitions between distant, secularly stable evolutionary branches. It has been shown that evolutionary gradualism and saltationism are complementary modes of macroevolution corresponding respectively to the local and global topographical structure of adaptive surfaces.

These results shall now be compared briefly with basic conclusions on speciation that Gould and Eldredge (1977) have derived from the fossil record. These authors believe that natural history consists in periods of morphological stasis interspersed with sudden bursts of speciation events. The latter tend to produce morphological discontinuities that are random with respect to the direction of a trend ("Wright's rule").

"Anagenesis is only accumulated cladogenesis filtered through the directing forces of species selection"

(Gould and Eldredge 1977, p. 141), where "species selection" means differential success of speciation events. Relative ability to speciate, not morphological flexibility, triggers evolutionary change.

It is evident that Gould and Eldredge's heavily debated view of organic evolution may become consistent with the features of phylogenetic succession in the present adaptive-topography picture combined with strong competitive selection, if it could be maintained that species-specific differences were necessarily discontinuous. Taking the congruence of discrete morphologies with actually or potentially interbreeding units as a tendency (rather than a rule), arising from the fact that the macrogenetic rearrangements in speciation tend to produce considerable effects on the phenotype, "Wright's rule" and "species selection" may be attributed to the non-local topographical properties of the adaptive surface. In particular, global anisotropies in the distribution of degenerate critical points across the surface may impose "directing forces" on "species selection."

As has repeatedly been stated by Ernst Mayr, and re-emphasised by Gould and Eldredge (1977, p. 117), species are normally characterised by optimal adaptation to their environments in the sense that they occupy ecological niches in which they prove superior to their competitors. This view being translated into the present adaptive-topography terminology, species correspond to bound regimes of phenotypic structural stability within which morphological homeostasis is maintained through competitive exclusion. "Relative ability to speciate" increases when populations evolve towards the boundaries of stasis regimes. It is an effect of structural complexity in polymorphic systems.

The results of the preceding investigation have been derived for competitive selection under the conditions of sympatric evolution. The significance of the analysis is expected to be less restricted, however. Consider a group of organisms becoming geographically and thus reproductively isolated from their ancestral population, so that allopatric speciation sets in. Evolution under the conditions of the new habitat then starts from the founder generation, which still shares some reasonable amount of genetic and phenotypic similarity with the ancestral population. Hence, the initial phase of allopatric speciation, at least, is always contaminated with elements of sympatric evolution to which the above considerations may apply.

It is interesting to note that the empirical evidence for discontinuous macroevolutionary phenomena, too, has motivated Gould and Lewontin's critique of the "adaptationist programme" (cf. Sect. IV.1). The present structural-instability approach to competitive selection and ecological adaptation demonstrates, however, that from the point of view of punctuational macroevolutionary change there is little justification for this critique. One may rather get the impression of another pointless scientific controversy which, according to Section III.5, may be resolved by suitably parametrising the controversial systems, laws and theories.

3 Dynamics and Structural Change in Biocultural Coevolution

Among the most intricate theoretical questions posed by evolutionary biology, one has the problem of how to explain the capacities of human self-organisation in evolutionary terms. It is widely accepted today among the scientific community that these capacities, which are usually referred to as *culture*, have evolved from less complex organic systems by the same genetic and ecological processes as other forms of life, too. However, the extreme variety of phenotypically distinct forms of human existence make it difficult to trace the phyletic pathways of human cultural capacity back to the prehuman forms of life. Some theorists even contend that the human phylogeny can be described best as a sequence of punctuational evolutionary changes (Gould and Eldredge 1977; Eldredge and Tattersall 1982) through each of which much genetic variance in phenotypic expressions was lost, eventually leaving *Homo sapiens* as a *tabula rasa* species with regard to genetic constraints on organic reaction ranges (environmental determinism).

Recently, Cavalli-Sforza and Feldman (1981), Lumsden and Wilson (1981) and Boyd and Richerson (1985) presented theoretical accounts of various aspects inherent in the evolution of cultural capacities in man. Lumsden and Wilson's book is especially concerned with the "coevolutionary circuit" through which genetic and cultural evolution drove each other forward, thus giving the impression of a self-accelerating process similar to the autocatalytic reactions known from biomolecular chemistry (Eigen and Schuster 1979; Babloyantz 1986). The mathematical framework of modern population genetics applied in these investigations proved a useful tool to treat analytically genetic fitness differentials arising from culturally codetermined, environmental and social interactions. A major difficulty in these approaches seems to be, however, to give the term "culture" an adequate operational definition which can be incorporated in the calculus, thus establishing the correspondence between abstract mathematical variables and specific cultural observables. For instance, the term "culturgen" (Lumsden and Wilson 1981), denoting cultural traits as objects of natural selection, has become subject to some criticism because of its alleged lack of realistic content (Leach 1981; Cloninger and Yokoyama 1981).

In the present approach to biocultural evolution, we concentrate on concepts of cultural ecology in order to avoid the numerous difficulties in assessing individually acquired and culturally transmitted traits in terms of *genotypic* fitness. There are a number of different schools in current anthropology focussing on population-environment interactions as major subprocesses of human biological and sociocultural development (for review and discussion see, e.g., Steward 1968; or Winterhalder 1980). The corresponding variety of theoretical perspectives and methodological approaches is matched by the common fundamental postulate that ecological factors constitute important causal agents of human biocultural evolution. In particular, White (1959, pp. 33–57), Adams (1975) and others largely identify the rates of matter and energy conversion in human ecological and social interactions with the rates of cultural development and differentiation (see also White 1954). It is the aim of the present section to put these aspects of biocultural evolution into the framework of parametrised dynamical systems.

The ranges of significance and validity of theories of cultural ecology have been subject to vigorous debate among contemporary anthropologists (for review and references see, e.g., Winterhalder 1980). Instead of dealing with the details of this controversy here, we develop basic aspects of the ecological approach to biocultural evolution in analytic terms, and test the results of the analysis against various empirical features of hominid evolution. Accordingly, we dispense with specifying culture in terms of human customs, symbols or social relations, and give an exclusively functional, dynamic characterisation of the notion of culture instead. Firstly, a simple mathematical model frequently considered in population dynamics (cf. Sect. V.2.2) is employed to simulate basic processes of organic evolution. Then an equation is established which describes the enhanced synthesis or organic matter by an intelligent tool-using species. The species is supposed to live partly upon these artificially reproduced organic resources. It is shown that the reproduction of both the species and its resources can be interpreted as autocatalytic reactions in the sense that the respective reproduction rates are increasing functions of the concentrations of the reactants involved. The rate constant entering the resource reproduction rate is referred to as the (overall) *cultural capacity* of the intelligent species. This definition is justified by intuitive arguments. Thus culture is characterised as a mode of ecological interaction. Finally, the evolution towards larger cultural capacities in a lineage of species with increasing intelligence and learning abilities is investigated. The heritable variation for cultural capacities is assumed to correlate with certain ranges of phenotypic variability within which differences in selective value arise from individually acquired traits transmissible through learning. There is thus a time lag between the occurrence of mutations conferring advanced capacities of self-organisation on individuals, and the development of novel phenotypes (rise and spread of cultural innovations). Hence effective mechanisms of maintained heritable polymorphism are necessary in order to exploit the hidden variation for cultural capacities.

The following analysis demonstrates that the ecological niches accessible to hominid species become broader and broader as the subsistence techniques improve, with releasing competitive selection pressure on the populations involved. This release favours the maintenance of polymorphism in these populations, which, in turn, extends the time span available for the exploitation of previously hidden adaptive potentials through individual learning and tradition. The latter step leads to further ecological release and thus closes the feedback loop in the "coevolutionary circuit".

3.1 The Basic Equations

In order to give a simple description of organic evolution in analytic terms (which is suitable to simulate evolutionary processes in secular time scales), a species or population S_1 is considered whose density X_1 obeys a Lotka-Volterra type of equation of logistic growth

$$\frac{dX_1}{dt} = k_1(N_1 - X_1)X_1 - dX_1 \qquad \text{(V.65)}$$

where the reproduction rate $k_1 N_1 > 0$ and death rate $d_1 > 0$, and the coefficient N_1 is proportional to the available amount of food and other resources. The species is supposed to live in a steady and homogeneous environment so that k_1, N_1 and d_1 can be treated as constants. Linear stability analysis of the stationary solution

$$\bar{X}_1 = N_1 - d_1/k_1 = K_1 \qquad \text{(V.66)}$$

shows that (V.65) is stable against small perturbations if the saturation level (or "carrying capacity") K_1 is positive.

Let a mutant subpopulation S_2 now appear in S_1 by genetic mutations or immigration, and let S_2 be characterised by population parameters k_2, N_2 and d_2 (all positive and constant). The mutants compete with S_1 for a share β ($0 \leq \beta \leq 1$) in the available resources. As in (V.53) to (V.57), the case $\beta = 1$ implies identical ecological niches of S_1 and S_2, whereas $\beta < 1$ indicates a partial overlap ($0 < \beta$) or complete separation ($\beta = 0$) in the respective resources. Then the competition between S_1 and S_2 is described by

$$\frac{dX_1}{dt} = k_1(N_1 - X_1 - \beta X_2)X_1 - d_1 X_1 \qquad \text{(V.67a)}$$

$$\frac{dX_2}{dt} = k_2(N_2 - X_2 - \beta X_1)X_2 - d_2 X_2. \qquad \text{(V.67b)}$$

The ecosystem (V.67) has been analysed briefly in the preceding section.

As has been pointed out by Nicholis and Prigogine (1977), the population dynamics (V.65) is isomorphic to the rate equation of an autocatalytic production of a chemical substance (concentration X) that interacts with a finite reservoir of another reactant A according to the reaction scheme

$$A + X \xrightarrow{k} 2X \qquad \text{(V.68a)}$$

$$X \xrightarrow{d} \text{inactivation} \qquad \text{(V.68b)}$$

where k and d are reactions constants. The chemical with concentration A, in turn, is supposed to be supplied at the same rate at which X is inactivated so that the overall concentration $X + A$ remains constant. Hence, this assumption may be referred to as the condition of quasi-closedness or, equivalently, dynamical equilibrium of the system composed of X and A. The corresponding rate equations then read

$$\frac{dX}{dt} = kAX - dX \tag{V.69a}$$

$$\left(\frac{dA}{dt}\right)_{\text{supply}} = dX \tag{V.69b}$$

$$\frac{dA}{dt} = -kAX + \left(\frac{dA}{dt}\right)_{\text{supply}} = -\frac{dX}{dt} \tag{V.69c}$$

the latter of which gives the conservation relation

$$A + X = N = \text{constant} \tag{V.70}$$

so that (V.69a) adopts the form (V.65).

The significance of the analogy between (V.65) and (V.68) for the present approach rests upon the conditions (V.69c) and (V.70) of quasi-closedness. Their application to the system composed of the species S_1 and the amount $A_1 = N_1 - X_1$ of its actual resources may serve in explicating the concept of adaptation. If in a steady and homogeneous environment

$$-\frac{dA_1}{dt} \neq \frac{dX_1}{dt}, \tag{V.71}$$

S_1 would either reproduce suboptimally or overexploit the reservoir A_1. For the remainder of the present analysis we therefore assume that all species concerned obey balance equations of the form (V.70) and are in this sense adaptive. The assumption implies that disequilibrium ecologies violating (V.70) become sooner or later subject to saturation phenomena re-establishing the dynamic preservation of the total amount of matter present ("mature", or "climax" stage of an ecosystem; see Odum 1969). Alternatively, it may be based on time averages over dynamic variations when the evolution of ecosystems in secular as against dynamic time scales is considered (Sect. V.2). For instance, in ecologies of the Lotka-Volterra type with fluctuating population numbers matter preservation indeed holds in time scales much larger than the Lotka-Volterra cycles (Goel et al. 1971).

The preceding considerations can be generalised straightforwardly so as to extend to ecological interactions that must be regarded as characteristic of human culture. Here, in a first approximation culture is understood as the capacity to manipulate purposely the environment and, more specifically, organic reproduction rates in human resource bases (Greenwood and Stini 1977, p. 400). This view, which reflects conceptions of cultural ecology as applied to human subsistence activities (Steward 1968, Netting 1971), is discussed in more detail below. Consider a species which has acquired the capability to use tools and specialises on the enhanced reproduction of its organic resources. This mode of ecological interaction can be described by a reaction scheme similar to (V.68),

$$R + A + X \xrightarrow{g} 2A + X, \tag{V.72a}$$

where the reaction rate is proportional to the density X of the species. The occurrence of X on both sides of (V.72a) shows X as an intermediate whose concentration remains unchanged, whereas A catalyses its own production. Furthermore, the reaction (V.72a) is supposed to be open to interaction with the reservoir R which furnishes the matter and energy necessary for the enhanced synthesis of A. If g in (V.72a) is sufficiently small, the exploitation of the reservoir is slow, and R may be treated as a constant. This approximation breaks down when we consider the case of large g below.

We further supplement (V.72a) by its reverse process

$$2A + X \xrightarrow{h} X + A + R, \tag{V.72b}$$

which involves two effects, firstly, an enhanced consumption and decomposition of A by X, and, secondly, the recycling of the decomposed organic matter A into the reservoir R. Under the conditions (V.69b) and R = constant, the rate equations read

$$\frac{dX}{dt} = kAX - dX \tag{V.73a}$$

$$\frac{dA}{dt} = -\frac{dX}{dt} + gARX - hXA^2. \tag{V.73b}$$

The existence of a non-trivial stationary solution requires

$$\bar{A} = d/k = gR/h = \text{constant}.$$

Choosing $gR = \alpha d$ and $h = \alpha k$, we get

$$\frac{dA}{dt} = -(1 + \alpha A)\frac{dX}{dt} \tag{V.74}$$

so that

$$X = N - \alpha^{-1} \ln(1 + \alpha A), \tag{V.75}$$

where N is an integration constant measuring the carrying capacity of a characteristic environment of X. Note that $\lim_{\alpha \to \infty} X(\alpha) = N$ and $\lim_{\alpha \to 0} X(\alpha) = N - A$.

The mathematical framework developed thus far is based on principles of competitive selection, with the coefficients N, k, d, etc. describing epigenetic interactions between species-specific phenotypes and environmental features. Now it is well known that neo-Darwinian synthetic theory requires finite genetic variance in phenotypic expressions exposed to natural selection, be the selection mechanisms frequency-dependent or density-dependent (Roughgarden 1979, p. 311). However, it is not necessary to make the genetic constraints on phenotype explicit when the selection of biocultural traits is considered below. It can be shown that the coefficients entering the governing equations are current average values of density-dependent genotypic fitness in polymorphic populations. The proof can be found in Roughgarden (1979, pp. 312–313) for a one-locus, diallelic system, and has been extended to multiple alleles for density-dependent selection, incomplete dominance and random mating in Section V.1. Hence the assumption of non-zero genetic variance in selected traits, on which the following analysis depends, is approximately, though implicitly, expressed by the present competition equations.

On the other hand, (V.73) differs from other theoretical approaches to competition and density-dependent selection that have been attempted in recent years (e.g. Roughgarden 1979; Ginzburg 1983). In our approach, selective advantages are evaluated in terms of genetically constrained capacities to manipulate resource reproduction rates which, in turn, control population growth. There is thus one step beyond (V.65) and (V.69) made explicit in the simulation of population-environment interactions, which leads to the cubic terms in (V.73) and (V.74), and higher-order non-linearities in the competition equations obtained below.

3.2 The Concept of Cultural Capacity

The biological evolution of the human capacities of social communication, co-operation and environmental interaction is clearly a multifactored process of extreme complexity. It involves such diverse phenomena as brain size and functional differentiation, learning abilities, tools, language, plasticity of social organisation and so on, and their multiple interplay (see, e.g., Wilson 1975). In the coevolutionary context, at least, these phenomena have to be treated from one common viewpoint, namely, the extent to which they confer on organisms the ability to extract as much matter and energy from their environments as is needed for survival and reproduction. Some biologists, anthropologists and sociologists seem to give less weight to this aspect by defining culture in terms of modes and devices of information processing other than genetic inheritance, that is, transmission of behaviours, skills and knowledge through teaching and individual learning (Bonner 1980; Cavalli-Sforza and Feldman 1981). On the other hand, cultural ecologists (Steward 1968; White 1959, Chap. 2; Netting 1971; Adams 1975) have insisted on the point that, when viewed functionally, the sociocultural capacities of prehistoric and modern man are subfunctions of certain processes of matter and energy conversion (cultivation, reproduction and recycling of natural resources) which constitute "the dynamics behind the entire cultural system" (Adams 1975, p. 281). However rough the conversion scheme (V. 72) may seem in comparison with real ecological interactions in hominid populations, it stimulates precisely these global ecological effects and thus retains an important component of the concepts of culture and cultivation. Similarly, the rate constant α entering (V. 74) and (V. 75) serves as an approximate measure of the overall effectiveness with which the species assumed in (V. 72) and (V. 73) reproduces, consumes and recycles its resources. Therefore, α may be termed the cultural capacity of this species which, on its part, may be regarded as a species of intelligent tool-using beings. This view indeed amounts to a modified restatement of Adams' (1975, p. 271) main thesis that "the rate of cultural change is proportional to the rate of energy conversion carried out within the system".

Reaction schemes similar to (V. 72) have also been used by White (1959, pp. 40–55) to analyse the connexions between culture, tools and rates of energy conversion in human environmental interactions. White maintains that culture can be understood as the human exosomatic, traditional organisation of customs, tools, language and beliefs which acts so as to exploit natural resources serving the needs of man. He extends the use of the term "tool" to cover all the material means employed by man to set conversion processes like (V. 72) into operation. Taking this view as a fundamental contribution to cultural evolutionism, we may indeed propose that the framework of (V. 73) retains relevant connotations of more broadly circumscribed concepts of culture used elsewhere in evolutionary anthropology. Only this restricted meaning of the term "culture" is intended when α is treated as the overall cultural capacity of a tool-using species here. Nor do we engage in the extensive anthropological debate on the question of whether all sociocultural phenomena in man can be reduced to (i.e., explained in terms of) cultural ecology, as some cultural ecologists seem to suggest (White 1959, p. 55; Adams 1975, p. 104). This question leads beyond the scope of the present section.

In the remainder of this section, the cultural capacity α as an evolving trait is investigated. Starting from $\alpha = 0$, a succession of species with increasing α and mean characteristic population size

$$\bar{X}(\alpha) = K(\alpha) = N - \alpha^{-1} \ln(1 + \alpha d/k) \tag{V.76}$$

is considered, whereby N, k, d and β are kept constant throughout. Evolution proceeds by invasion of previously existing populations with $\alpha = \alpha_1$ by mutants carrying the cultural capacity $\alpha = \alpha_2$ so that competition sets in (cf. (V.67), (V.69) and (V.75)),

$$\frac{dX(\alpha_1)}{dt} = k\alpha_1^{-1} X(\alpha_1) \exp(\alpha_1(N - X(\alpha_1) - \beta X(\alpha_2)))$$
$$- (k\alpha_1^{-1} + d) X(\alpha_1) \tag{V.77a}$$

$$\frac{dX(\alpha_2)}{dt} = k\alpha_2^{-1} X(\alpha_2) \exp(\alpha_2(N - X(\alpha_2) - \beta X(\alpha_1)))$$
$$- (k\alpha_2^{-1} + d) X(\alpha_2). \tag{V.77b}$$

The obvious assumption is made that the control parameters (as opposed to the dynamic variables) N, k, d, β and α are determined by epigenetic interactions between the genome and characteristic environment of a population, thus specifying reproductive differences among species-specific phenotypes. We further assume long-term random genomic substitutions and environmental transformations in time scales τ much larger than the inverse relaxation rate of (V.77) which will, in general, change the control parameters ("evolution in secular time scales"). However, we only consider variations in α here. This restriction will be discussed in the next paragraph. The stability analysis of (V.77) proceeds exactly as outlined in the previous section and yields results analogous to (V.53) to (V.57). In these relations, one must only substitute K_1 and K_2 by $K(\alpha_1)$ and $K(\alpha_2)$ respectively. Choosing $\beta = 1$ (which will be justified below), the evolution of α satisfies

$$\frac{dK(\alpha)}{d\tau} = \frac{d\bar{X}(\alpha)}{d\tau} > 0 \tag{V.78a}$$

This result follows straightforwardly from the competitive structure and stability behaviour of the system (V.77), which guarantees that only those mutant phenotypes will be successful which are characterised by

$$\frac{d\alpha}{d\tau} = f(\tau) > 0. \tag{V.78b}$$

One thus gets an evolutionary sequence of stable population equilibria with optimal degrees of adaptation evaluated in terms of $K(\alpha)$.

Several qualifications have to be made concerning the reality of the assumptions adopted so far. From our present knowledge of the rise of cultural traits in the hominid line (e.g., Gowlett 1984), it is obvious that biocultural evolution has had a strong impact on all conceivable population parameters. Hence, in the present equations the birth and death coefficients should be treated as variables in α as well. Since changes in these parameters are quite general features of organic evolution, however, we neglect these effects here and concentrate exclusively on the investigation of (V.72), which covers basic characteristics of the notion of culture. Similarly, one must expect that the richness of the ecological niches accessible to hominid populations in natural history were strongly dependent on the technological skills of these populations such that a function $N(\alpha)$ in (V.75) to (V.77) might seem more appropriate than $N = $ constant. However, the impact of advanced subsistence techniques on the amount of available resources has already been taken into account. Expressed in terms of α, (V.69b) reads

$$\left(\frac{dA(\alpha)}{dt}\right)_{\text{supply}} = dX(\alpha). \qquad (V.79)$$

This means that larger α and $X(\alpha)$ require an increase in supply rate which (V.72) cannot provide; it depends on the acquisition of new resources. Furthermore, (V.79) may serve for estimating the degree β to which the ecological niches of two species with α_1 and $\alpha_2 > \alpha_1$ overlap:

$$\beta \cong \frac{(dA(\alpha_1)/dt)_{\text{supply}}}{(dA(\alpha_2)/dt)_{\text{supply}}} = \frac{X(\alpha_1)}{X(\alpha_2)}.$$

As we consider only continuous evolutionary changes in α here, the "local" approximations $\alpha_2 \gtrsim \alpha_1$ and $\beta \cong 1$ always suffice.

The system (V.72) suggests an interpretation as a schematic ecology of the agrarian type in which manipulations of resource reproduction rates (tillage and stock-breeding; see (V.72a)) and intensified fertilisation (recycling; see (V.72b)) are the prevailing modes of environmental interaction. However, agricultural systems have appeared very recently in human natural history and, hence, could not have had much impact on the processes of organic differentiation and natural selection. Nonetheless, effects of cultural ecology similar to agriculture were already operating in earlier hominid populations. For instance, the use of fire as an artificial extraction of heat from organic matter acts like (V.72) if heat is included among the natural resources A. The primary source of energy for all organic life is solar radiation catalysing the production of plant tissue from which, in turn, heat can be extracted by fire techniques. Thus the use of fire can be understood straightforwardly as an artificially enhanced production and consumption of energy exercised by tool-using intelligent beings. Another example of pre-agrarian forms of cultivation and intensified resource reproduction is given by interspecific aggression and competition. Deliberately killing other predators preying on the same game species as the early hunter-gatherers (Wilson 1975) acts so as to increase the rate constant in (V.72). Finally, slash-and-burn ecological strategies, ethnographically recorded among hunter-gatherers, are probably the oldest techniques to increase the net primary productivity of human ecosystems (Stewart 1956). Other examples of the reproduction and recycling of biomass and the harnessing of the chemical energy stored in it have been extensively dealt with by White (1959, Chap. 2) under the rubric of "energy and tools".

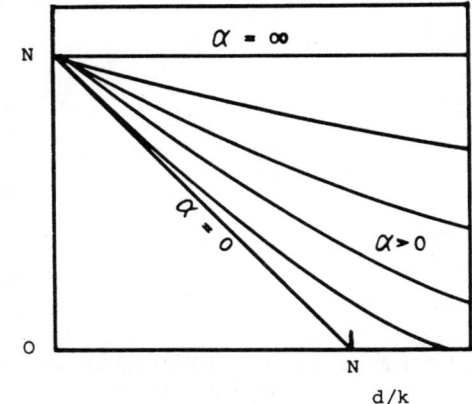

Fig. V.6. Population number $\bar{X}$ as a function of the ratio d/k for various values of the overall cultural capacity α. For $\alpha > 0$ the *curved lines* show that $\bar{X}$ remains finite in ecological niches with $d/k > N$, where populations at zero cultural capacity could not exist (after Geiger 1985a)

3.3 The Impact of Learning on Biocultural Evolution

In Fig. V.6 we have schematically drawn $\bar{X}$ as given by (V.76) as a function of the ratio d/k for various values of α and for an imaginary environment of the carrying capacity $K = N - d/k$ in dynamic equilibrium. Various ecological niches are distributed over the environment which co-determine the equilibrium values $\bar{A} = d/k$. The straight line $\alpha = 0$ demarcates the set of niches with $0 \le d/k \le N$ whose carrying capacity has the absolute upper boundary N fixed by environmental "limiting factors" (Odum 1969). The curved lines give $\bar{X}$ as a function of d/k for $\alpha > 0$. $\bar{X}(\alpha)$ is not only increased above $\bar{X}(0)$, but also remains positive for $d/k > N$. Hence species with $\alpha > 0$ may penetrate ecological niches where they could not exist otherwise, that is, at zero cultural capacity. In the asymptotic case $\alpha \to \infty$ the limiting carrying capacity N of the conceived environment is made use of by a highly skilled species which transforms the total inflow of organic matter and other resources into optimal population size according to (V.79): $\lim_{\alpha \to \infty} (dA/dt)_{\text{supply}} = dN$.

It is obvious from the preceding considerations that evolution towards higher α obeys the principle of competitive exclusion. As α approaches infinity, however, competitive selection pressure between successive species operating at higher and higher α is released. This is illustrated in Fig. V.7 where we have plotted $\bar{X}$ as a function of α for fixed d/k. At low α, fluctuations $\Delta \alpha$ yield large increments, and, for negative $\Delta \alpha$, decrements $\Delta \bar{X}$ in population size, whereas $d\bar{X}/d\alpha \to 0$ for $\alpha \to \infty$. Linear stability analysis of (V.77) shows that the rate per $\Delta \alpha$ at which (V.77) relaxes to a new stationary state after the variation $\Delta \alpha$ occurred is given by

$$\omega = (k + \alpha d)\frac{d\bar{X}(\alpha)}{d\alpha}, \qquad (\text{V.80})$$

so that for the mean period ω^{-1} of maintained polymorphism between the unperturbed and the mutant subpopulations we get

$$\lim_{\alpha \to \infty} \omega^{-1} = \lim_{\alpha \to \infty} \left[(k + \alpha d)\frac{d\bar{X}}{d\alpha}\right]^{-1} = \infty. \qquad (\text{V.81})$$

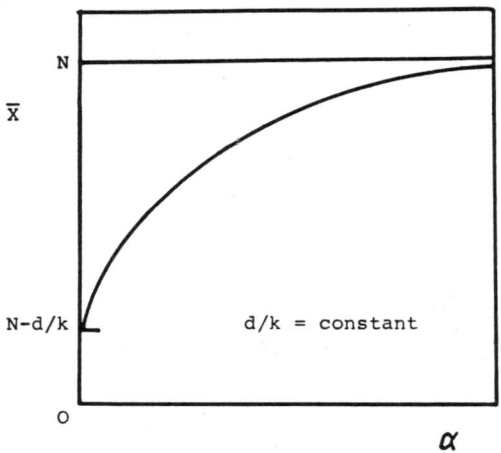

Fig. V.7. Population number $\bar{X}(\alpha)$ for fixed ratio d/k and biotic capacity N (after Geiger 1985a)

If the present model correctly simulates evolutionary mechanisms specific of the hominid phylogeny, it predicts a tendency towards competitive exclusion between the early hominid species competing for their habitats, whereas coexistence becomes more likely as biocultural evolution proceeds. Unfortunately, the present fossil evidence still seems to be too inconclusive to support or falsify this proposition empirically (for review and discussion of this question, see McEachron and Baer 1982, pp. 135–136).

The present assumption that there is finite genetic variance in phenotypic expressions exposed to natural selection is made somewhat more explicit now. We suppose in first approximation that there is a steady source of genetic variation which leads to $f(\tau) = \mu =$ constant in (V.78). If the genetic variance in α were always very high, (V.78) would yield

$$\frac{d\bar{X}}{d\tau} = \frac{d\bar{X}}{d\alpha}\frac{d\alpha}{d\tau} = \mu \frac{d\bar{X}}{d\alpha}. \tag{V.82}$$

This relation is schematically depicted in Fig. V.8.

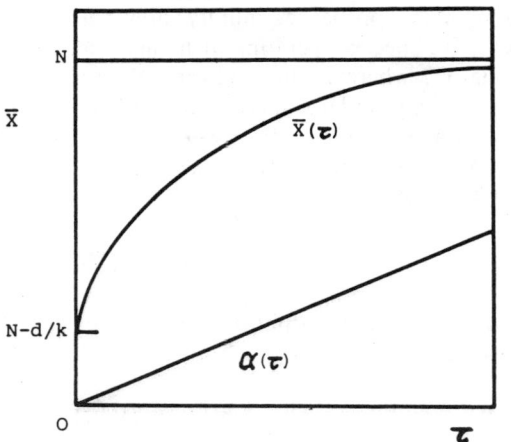

Fig. V.8. The evolution of the cultural trait α and mean characteristic population size $\bar{X}$ over secular time scales τ for constant average rate of evolution and complete genetic determination of α (after Geiger 1985a)

The situation changes when individual learning and tradition contribute to the specification of α. The presence of variation for α then means a certain disposition to learning in individuals, and reproductive fitness becomes a function of individually acquired traits. Correspondingly, increases in cultural capacity tend to arise from hidden adaptive potentials which will not yield selective advantages in time scales shorter than one generation (as an absolute lower boundary). The time per $\Delta\alpha$ between a mutation and the appearance of a new phenotype through learning ($\alpha \to \alpha + \Delta\alpha$) is of the order of magnitude $T = n(kN)^{-1}$, where $(kN)^{-1}$ is of the order of magnitude of one generation and n is a positive integer which will generally be large. Effective mechanisms of maintained hidden variation thus become codeterminants of the evolutionary success of mutants which will express higher values of α after the time T if they are not counterselected immediately after their first appearance. These effects give a modified evolutionary rate (cf. (V.80))

$$\frac{d\alpha}{d\tau} = \frac{\mu}{\omega T} = \mu N \left[n(1 + \alpha d/k)\frac{d\bar{X}}{d\alpha} \right]^{-1}, \tag{V.83}$$

where $(\omega T)^{-1}$ is the ratio of the time scales of the maintenance of non-stationary fluctuations $\Delta\alpha$ around α, and of learning and cultural transmission. For small α one has $(\omega T)^{-1} \ll 1$. In this case mutants are exposed to strong competitive selection, and the large relaxation rates ω of the system (V.77) prevent the development of traits expressing high individual flexibility. For large α, when $(\omega T)^{-1} \gtrsim 1$, competitive selection pressure on mutants eases. Every mutation contributing to an increase in α becomes more likely to carry through its selective advantage without restriction, and the genetic variance in phenotypic expressions is eventually reduced. The long-term evolution of α then obeys (V.83) and

$$\frac{d\bar{X}}{d\tau} = \frac{d\bar{X}}{d\alpha}\frac{d\alpha}{d\tau} = \frac{\mu N}{n(1 + \alpha d/k)} \tag{V.84}$$

as sketched in Fig. V.9. The steep quasi-exponential rise of α may be compared with certain conspicuous gross features of hominid evolution such as the accelerated increase in brain size (Pilbeam 1972) which it parallels qualitatively. On the other hand, the present analysis predicts a very slow increase in mean population size. This, however, is consistent with the data and semi-empirical hypotheses on overall Pleistocene population growth which has been estimated at about 0.001% or less per year, and probably did not vary considerably during that period (Cohen 1980; Hassan 1980).

3.4 The Expiration of the Coevolutionary Circuit

As noted above, in the limit $\alpha \to \infty$ the basic assumptions of the ecosystem (V.72) break down. The secondary production rate at which A is synthesised at the expense of the matter and energy resources R is clearly limited by the rates of biological growth and decay, and of environmental processes such as photosynthesis, ecological succession (Odum 1969) or seasonal variation in climate and weather. Increases in α beyond these limitations can, at best, be attained by intensified depletion of the primary resources R whose concentration can then no longer be treated as a constant. One must therefore expect that somewhere on the steep branch of the α-curve in

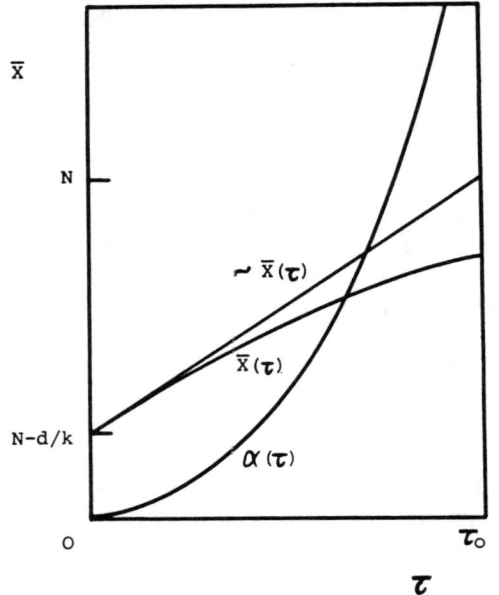

Fig. V.9. The evolution of the trait α and average population size $\bar{X}$ over secular time scales τ. It is assumed that α is codetermined by individual learning and cultural tradition. The *straight line* labelled $\sim \bar{X}(\tau)$ represents a rough approximation to $\bar{X}(\tau)$, with the factor $(1 + \alpha d/k)^{-1}$ neglected in (V.84) (redrawn from Geiger 1985a)

Fig. V.9 a transition to intrinsically non-stationary ecosystems occurs when depletion and consumption of natural resources proceed faster than biomass can be refurnished by biological processes in dynamic equilibrium. In fact, it is well known that the rise and spread of advanced agriculture and intense forestry in history shifted the human ecosphere away from the "mature" equilibrium states in which the ratio A'/R' of the "standing crop biomass" A' (as a part of A) and the amount R' of primary plant tissue acting as a reservoir of nutrients (as a part of R) adopts some stable optimum value. Agrarian systems moved more and more towards unbalanced ecologies in which the ratio A'/R' is suboptimal, implying an ever higher consumption of the primary matter and energy resources R (Odum 1969).

In the present account, the evolution of large cultural capacities, the expiration of competitive selection, and the rise of intrinsically unbalanced ecologies virtually coincide. This is expected to happen when the time scales T of the spread of technological innovations ("history of civilisation") become much shorter than those of competitive selection ($\omega^{-1} \gg T$). Our approach admits an order-of-magnitude estimate of the time τ_0 of the onset of the history of civilisation when the factor $(1 + \alpha d/k)^{-1}$ is omitted in (V.84). In this approximation, the integration of (V.84) from $\tau = 0$, $\bar{X}(0) = N - d/k$ to $\tau = \tau_0$, $\bar{X}(\tau_0) \simeq N$,

$$\tau_0 \simeq \frac{nd}{\mu k N} \tag{V.85}$$

shows that τ_0 is determined essentially by the inverse average rate μ^{-1} of evolution. Since in our representation μ is not simply proportional to the rate of genomic substitution (which can be estimated; see, e.g., Mayo 1983; Nei 1987) but also covers the complicated transformation of genetic dispositions into phenotypic expressions, it may be determined indirectly by inserting the estimated Pleistocene population growth rate of 10^{-5} year^{-1} for $d\bar{X}/d\tau$ in (V.84) so that $\mu \simeq 10^{-5} nN^{-1}$ year^{-1}. Relation (V.85) then yields

$$\tau_0 \simeq 10^5 \, (d/k) \text{ years} \simeq 10^5 \, N \text{ years},$$

where $d/k = N - \bar{X}(0) \simeq N$ by order of magnitude. An appropriate chocie for N would be the typical size of local hunter-gatherer groups extrapolated backward from living primitive societies. Such extrapolations have frequently been carried out in the literature (Wilson 1975; Cohen 1980; Hassan 1980) with the widely agreed result of $N \simeq 10^2$. These estimates leave τ_0 of the order of some million years which roughly covers the Pliocene-Pleistocene period. Hence, within the approximations on which the present approach is based the empirical time scales of hominid evolution are reproducible numerically.

3.5 Summary and Conclusions

A simple mechanism of biocultural evolution has been proposed which is based on three obvious assumptions. Firstly, phylogenetically successive species are exposed to competitive selection. Secondly, culture is characterised as a mode of ecological interaction arising from the intelligent manipulation of natural resources, and exercised by tool-using species. Thirdly, the capacities of individual learning and tradition are organic traits subject to biological constraints. Once these conceptions are translated into an appropriate analytic framework, a variety of empirical features of hominid evolution can be explained in a straightforward manner. The approach correctly predicts the widening range of accessible ecological niches and the decrease in immediate genetic impact on phenotype as biocultural evolution proceeds. The essence of the present approach, however, is expressed in Fig. V.9, which reflects two fundamental characteristics of hominid evolution, namely the self-accelerating increase in technological capacity, with characteristic population size varying much less drastically. The former phenomenon is explained by a kind of autocatalysis mechanism (Hart 1959; Wilson 1975, pp. 566–568; Stebbins 1982), meaning that every successful evolutionary step towards large cultural capacity provides selective advantages for prospective moves in the same direction. The latter is clearly a saturation effect (limits imposed on Malthusian growth by the condition of ecological equilibrium).

The impact of ecological constraints on human prehistoric population regulation is still a matter of controversy (see various contributions to the book edited by Cohen et al. (1980), and Hayden (1981)). In particular, Hayden has questioned the empirical significance of carrying capacities and other factors related to ecological equilibrium, referring to constantly fluctuating environmental conditions throughout man's natural history. Commentators on Hayden's conclusions such as Cohen (1981) and Druss (1981) have argued, however, that periodic variations in the resource bases of early man must be expected to modify the effects of ecological saturation rather than cancel them completely. Similar arguments can be drawn from the mathematical analysis of ecological cycling which demonstrates that time averages of fluctuating population numbers and resources may simulate the presence of population equilibria in secular time scales (Goel et al. 1971; see also previous section). Now on the basis of (V 75), the present approach suggests that as long as the overall saturation level N is invariant, the population size X remains bounded even for moderate α, since $X(\alpha)$ depends exclusively on the resource supply rate according to (V.79). Correspondingly, rapid population growth as is to be observed since the Neolithic may have been possible only due to marked increases (i.e., non-cyclic changes) in N, which are characteristic of inherently unsteady ecosystems, however.

VI Applications to Human Social Structure

1 The Anthropological Significance of Evolutionary Stability and Instability

So far theoretical extensions of the adaptive hypotheses of sociobiology have been considered. These extensions have been based on suitable parametrisations of the governing equations of mathematical population genetics and population dynamics. A unifying theoretical approach to both adaptive and non-adaptive evolution has become possible this way, and, more specifically, some long-term evolutionary effects have become linked to the mathematical concepts of evolutionary stability and instability in a systematic fashion.

We proceed to outline applications of the conceptual framework of evolutionary or, equivalently, structural stability to human social structure. Such applications are desirable for reasons which in the sociobiology debate of the past two decades have been subsumed under the notion that non-adaptive social behaviour is prominent in man. Since all the details of the debate are reported in the literature on sociobiology, we need not review them at length here. We simply refer to Boyd and Richerson's (1985) book for such a review, and for a most careful and penetrating theoretical analysis of the processes of human adaptive and non-adaptive evolution. The book provides an excellent example of the structure and utility of unified scientific theories (cf. Boyd and Richerson 1985, pp. 289–290), while the authors do not specifically follow the approach described in Part One. The concept of adaptation still being understood strictly in its Darwinian sense, we shall present only a few general considerations concerning non-adaptive human behaviour as a motivation here, and shall otherwise make the relevant points in those contexts of our analysis into which they belong. Recall, however, that when we start from the observation of non-adaptive human social behaviour patterns, we do not deny *any* anthropological significance of sociobiology at all. Once more we aim at biological perspectives complementary rather than alternative to the neo-Darwinian account of biosocial evolution.

Sociobiologists and committed anti-reductionists both admit that the characteristic time scales of human cultural history are much shorter than those of natural selection (see Sect. V.3.4) and that, therefore, the cultural and ecological constraints on human behaviour change faster than differential genetic transmission can effect behavioural adaptations. It is also widely agreed upon that humans possess biologically advantageous capacities to develop or maintain, in a variable environment, the locally optimal phenotype through learning and cultural transmission. Current sociobiological controversies are rather concerned with the problem of whether this optimisation proceeds so as to increase the genetic fitness of human social actors, i.e., whether "culture simulates biological adaptations" or not.

Suggesting a unified theoretical treatment of the issues raised in the debate, we take the changing ecological, cultural and historical constraints on human social behaviour as a structural-stability problem. In other words, we propose that human interaction patterns such as dominance, ritual and kin altruism with marked historical and cross-cultural invariance properties are structurally stable in the sense of population dynamics (Chaps. IV, V). Conversely, we also try to show that there exist gross modes of cultural evolution which can be theoretically understood as bifurcations arising in evolutionarily unstable human systems.

We shall argue that the evolutionary stability of human social systems draws upon phylogenetically acquired dispositions of social action which, in cultural contexts, have become functionally non-specific, however. In human history they persist as multifunctional behavioural dispositions of social organisations, suitable for serving a diversity of culturally contingent needs of individuals and groups. As is the case with any other instrument of human organisation, this suitability is constrained, however. The physical and psychophysiological correlates of behaviour confer different degrees of effectiveness on alternative modes of social action. Thus the goals of human social actors may well be culturally contingent, while the behavioural dispositions to realise these goals are biologically circumscribed. Accordingly, human biosocial relations may not only prove biologically adaptive, but may also indicate forms of rational social organisation and decision-making. At any rate, they will be anything but functionally arbitrary, as some anthropologist opponents to human sociobiology would insinuate (e.g., Sahlins 1976; Harris 1979). The latter simply mistake biological multifunctionality of an organic trait for functional arbitrariness. The present instrumentalist interpretation, or "biotechnology view", of human sociobiology should not be confounded with this kind of anthropological anti-adaptationism.

Before elaborating upon these conceptions, we illustrate them briefly by a key example drawn from the sociobiology debate. Sociobiologists are accumulating data suggesting that a variety of kinship classificatory systems, anthropologically recorded among cultures all over the world, are indeed adaptive in the sense of inclusive-fitness theory (Alexander 1979). This theory implies that the net benefit of unbalanced social reciprocation to inclusive fitness varies with the degrees of genetic relatedness and environmental conditions and, hence, is expected to account for the observed diversity of kinship classifications (Fox 1979). Some anthropologists strongly oppose this view. They stress the fact that in many societies kinship categories have merely a symbolic function and do not coincide with genealogical relationships (Sahlins 1976) which, on their part, can be redefined in terms of genetic relatedness. In the opinion of these anthropologists hypotheses on the adaptive significance of differences in human kinship systems are difficult to generalise.

Now the conceptual framework of structural stability suggests another interpretation. In the known diversity of kinship systems, basic recurrent patterns of altruism and social reciprocation can be observed (Essock-Vitale and McGuire 1980), indicating biobehavioural preadaptations of small-scale social organisation. These patterns provide organisational advantages regarding the satisfaction of a variety of proximate needs of human small-scale groups, be these groups organised according to genetic or whatever symbolic criteria of kinship. In Sahlins' (1976, p. 57) words,

"...kinship relations govern the real processes of co-operation in production, property, mutual aid and marital exchange..."

in small-scale societies. The situation described by Sahlins amounts to a reorganisation of the cognitive, developmental and physiological correlates of kinship when symbolic constraints are imposed, in culture-specific ways, on inclusive-fitness maximising social traits. This reorganisation, although non-adaptive in the strict neo-Darwinian sense, is not arbitrary, however. Social interactions in terms of symbolic kinship function so as to exploit biobehavioural dispositions to co-operate. Organising "the real processes of co-operation in production, property, mutual aid and marital exchange" in terms of symbolic kinship is a kind of biotechnology of cross-culturally optimal usefulness in sustaining small-scale social groups. The optimality properties of this mode of social organisation are accessible to comparative functional analysis.

The following two sections are especially concerned with a combined functional and structural-stability analysis of problems of increasing hierarchical complexity in cultural history. They include materials previously published elsewhere (Geiger 1985b, 1988c).

2 Political Power as an Evolutionary Structure

In recent years evolutionary biologists and social anthropologists have shown themselves increasingly interested in a unified theoretical treatment of human biological and sociocultural evolution. This situation is by no means a matter of course. Although E. O. Wilson (1975) intended such a unified treatment explicitly when incorporating anthropology in his evolutionary synthesis, his seminal book on sociobiology first had some quite adverse effects. It launched numerous scientific controversies in which the fundamentals of anthropology were reviewed and discussed in rather antithetical terms such as nature versus culture (Sahlins 1976, Part I, esp. pp. 11–13), genetic determinism versus environmental relativism (Caplan 1978, Part III; Montagu 1980), ultimate versus proximate causation of behaviour (Losco 1981), holism versus reductionism (cf. Rose et al. 1984, Chap. 10), and related conceptual dichotomies. However, since from the perspectives of biocultural co-evolution there is little point in rigidly disjunctive views of human nature and culture, evolutionary anthropologists seem to accept and assimilate increasingly empirical data, concepts and theories from ethology, behavioural physiology and sociobiology. In particular, the recent theoretical interest in biosocial and biocultural syntheses is largely due to demonstrations of marked patterns of inclusive-fitness maximising behaviour in primitive as well as complex societies (Alexander 1979; Chagnon and Irons 1979). There is now also vast literature treating differential transmission (selection, drift) of genetically and culturally heritable variation in phenotype as a universal evolutionary mechanism (e.g., Cavalli-Sforza and Feldman 1981; Boyd and Richerson 1985; Findlay et al. 1989). And in his elegant adaptation of evolutionary game theory, Axelrod (1984) pointed out the way in which patterns of co-operation may spread and lead to stable social structures in groups of egoistically acting individuals.

From a still broader theoretical perspective, these evolutionary, biobehavioural accounts of human social relations remain inherently incomplete, however. The situation can be circumscribed as follows. One of the most prominent features of cultural history since the Neolithic is the hierarchical differentiation, or increasing organisational complexity, of human social groups. In the modern life sciences, "hierarchical differentiation" is largely synonymous with "natural self-organisation", which roughly means the rise of organic macrostructures from the microphysical structures of prebiotic matter (see Sect. II.2). Now the historic rise of large-scale hierarchically differentiated societies from the microstructures of human face-to-face interactions (Fried 1967; Service 1975; Cohen 1978; Cohen and Service 1978) shares many theoretical aspects and problems with natural self-organisation. Problems in point are the stability and control of large-scale, multicomponent systems, and the processing and dissipation of energy and information within them. The common features of natural and sociocultural self-organisation become apparent, for example, from anthropological studies which link the processing of energy (Adams 1975) and information (Flannery 1972) in human systems with the evolution of complex society. But with a few exceptions (e.g., Willhoite 1981; Corning 1983), the issue of increasing hierarchical complexity has widely been neglected in *biobehavioural* approaches to human social structure. This neglect reflects the well-known fact that evolution in the sense of the neo-Darwinian concepts of natural selection and adaptation, which also frame sociobiology conceptually, is not consistently correlated with hierarchical differentiation (Williams 1974, Chap. 2; Corning 1983, p. 64). Nonetheless, increasing organisational complexity is a most frequent concomitant of organic, including sociocultural, evolution. It is therefore hard to conceive any unified theory of biosocial evolution which does not, in one sense or the other, encompass the issue of natural self-organisation.

The present chapter is designed as an evolutionary approach to human social structure and stability, combining biobehavioural concepts with notions of system complexity. It attempts to account for two basic empirical facts consistent with the anthropological data and current theories on state origin: Firstly, political organisation is a virtually universal characteristic of complex, as compared to primitive and ranked, society; secondly, the historic rise of the state and similar forms of political organisation cannot generally be understood as an adaptation process in the sense of sociobiology theory. The account concentrates on the function of political power for the maintenance and control of large-scale, hierarchically differentiated corporate groups. This function will be analysed and explained in biobehavioural terms.

The issue of social control in large-scale human groups is closely connected with the classical question of how to establish co-operation among selfishly motivated individuals (Hobbes, *Leviathan*, Parts I, II). This question has recently been reconsidered in terms of the neo-Darwinian evolution of social behaviour (inclusive-fitness theory, evolutionary game theory), with a number of important results (Trivers 1981; Masters 1983; Axelrod 1984). In particular, the evolution of co-operation has been demonstrated to require theoretically neither the presence of a central Hobbesian authority nor the subjective rationality of human social action. However, the assumptions on which these results have been obtained can reasonably be argued to neglect various realistic factors such as the discrepancies between adaptive evolution and hierarchical differentiation, and the ecological constraints of limited resources on co-operative structures (Masters 1983; Geiger 1985c). So basic theoretical questions remain open which call for an evolutionary account of human complex society leading beyond the domains of neo-Darwinian theory.

In the following sections we first briefly review and explain certain concepts and theories of the structural stability and instability of evolving systems that have been successfully applied to problems of natural self-organisation in physics and the life sciences. Then we show that the conceptual framework of structural stability/instability can also be employed in evolutionary explanations of complex sociocultural relations. It is argued that human complex society exhibits an intrinsic tendency to disintegrate which it shares with other living systems and which is characteristic of complex hierarchical organisation. This tendency leaves the evolution and maintenance of organic, including sociocultural, complexity a highly non-trivial problem which involves not only the adaptivity of systems with respect to environmental change but also control of the internal order and stability of hierarchical arrangements. Using the notions of political power and political organisation once set forth by Max Weber, it is argued that complex society owes its coherence and historical significance to a very peculiar power structure, namely the cultural transformation of certain universally stable organismic strategies of social control. Finally the results of the present investigation are contrasted with previous biobehavioural approaches to political organisation. It is concluded that, from the biosocial point of view, political organisation is characterised not so much by its usefulness in maximising the inclusive fitness of co-operating individuals but by its effectiveness in optimising social behaviour control.

2.1 Evolution of Hierarchical Complexity

The concepts of hierarchical complexity used in the natural and social sciences evidently refer to systems of quite different kinds which, however, share basic organisational features. Although the present analysis is largely concerned with multilevel systems of social control and decision-making that are usually referred to as *dominance hierarchies, social hierarchies* or *control hierarchies*, we use the term "hierarchy" exclusively to distinguish levels of description in compound systems. Masarović et al. (1970, Chap. 2) have pointed out numerous close connexions between the appropriate levels of conceptualisation, on the one hand, and the levels of control, on the other hand, in complex biological and socioeconomic systems. However, in order to prevent confusion, we avoid phrases such as "control hierarchy" altogether. The notion of hierarchy accordingly implies a distinction between microscopic levels of organisation in a system, or object of concern, while "complexity" means large numbers of constituent parts (microsystems) of any such object (macrosystem), the parts being highly interconnected among each other. Familiar examples of complex hierarchies are the successive community levels in ecosystems, each community comprising several interacting populations, with the populations consisting of large numbers of interbreeding and otherwise interacting individuals. At the microscopic level of conceptualisation, the behaviour of complex hierarchical systems is thus always characterised by many parameters (i.e., modes of motion, action, etc. of the constituent parts) even if global descriptions of the systems can be given in terms of a few macroscopic variables. As a statistical effect, the behaviour of the individual elements of a complex system shows random deviations from the average pattern characteristic of the macrolevel of organisation. Note that here "randomness" means that at the microlevel of organisation the possible modes of system behaviour constitute random variables in the observer's conceptual distinction rather than in an absolute sense. For example, if a biological population is described in terms of birth and death rates, the birth or death of an individual organism is a random event in this description although the event may be well determined from the point of view of developmental biology.

These considerations raise two fundamental questions concerning the ranges and limits of natural self-organisation. (1) The problem of the evolution of hierarchical complexity: How can large-scale coherent structures arise from previously uncorrelated motions, behaviours, etc. of huge numbers of individuals? (2) The problem of the stability of hierarchical complexity: Under which conditions are the statistical variations in microscopic behaviour kept sufficiently small so as to preserve the macroscopic order of complex hierarchical systems? The two problems are critical for complex systems the behaviour of which is *dynamic*, that is, depends sensitively on environmental constraints such as inflow of energy and matter, and generally varies with time. For this type of complex system the stability problem is indeed vital since even small random changes at the microlevel of organisation may show large-scale dynamical effects, for example, an initially small number of spontaneously appearing mutant genotypes may introduce entirely novel modes of population interaction into an ecosystem (cf. Allen 1976).

Generally, the evolution and stability of complex systems require theoretical approaches which combine elaborate stochastic theories of microscopic behaviour with dynamical theories of large-scale organisational phenomena (Nicolis and Prigogine 1977; Haken 1978, 1983). Useful informal treatments of the basic mechanisms of natural self-organisation have been given by Prigogine (1980), Prigogine and Stengers (1980) and Haken (1981). Here we concentrate on an approximate method of analysis which makes no explicit reference to microscopic behaviour and which has been described above as the structural-stability approach to dynamical systems (Hirsch and Smale 1974; Thom 1975; Poston and Stewart 1978).

The approach can be roughly described as follows. The macroscopic attributes of hierarchical systems are characterised in quantitative terms by suitably chosen variables which depend on each other and on time. Generally, there are several modes of interdependence of (i.e., dynamics for) these variables, each mode depending on internal constraints and environmental conditions. As an example, consider an ecosystem of interacting species. In this case, the dynamics is given by some mode of symbiosis or competition, the population densities are the dynamic variables, and the birth and death rates, trophic networks and energy inflows constitute internal and external constraints. If now the relationship between the constraints and the modes of dynamic behaviour is known, one can predict the extent to which changes in the constraints and environmental conditions will affect the global state and operation of the system. This possibility is expressed by the following definition, which we restate after Section IV.2 in an information fashion. A system is called *structurally unstable* if it responds to small changes in its constraints by transitions into qualitatively different modes of behaviour and organisation, and *structurally stable* otherwise. In particular, the formation and decomposition of complex hierarchical systems can be treated as internally or externally induced structural instabilities; for structurally unstable systems are driven into new regimes of ordered or disordered behaviour, respectively, if their macroscopic constraints show non-zero statistical variance. This will inevitably be the case if the systems are sufficiently complex so that random effects at the microlevel of organisation gain critical influence.

The concepts of structural stability and instability provide technical and theoretical advantages for studies in natural self-organisation. Structural-stability analyses dispense with all the technical difficulties of the stochastic approach to complex self-regulating systems, while retaining essential features of macroscopic order and evolution. Since the conceptual framework of structural instability refers to transitions between the different regimes of dynamic behaviour within which a system may be found, the complementary evolutionary effects of dynamic change and structural transformation can be subject to a unified theoretical treatment within this framework. Examples which combine the dynamics of natural selection and adaptation with effects of structural instability in populations and ecosystems have been treated in Chapter V. As for recent mathematical approaches to the dynamics of biocultural systems, see Findlay et al. (1989).

2.2 Stability of Social Structure

Unfortunately, it is very difficult to determine, in theoretically satisfying ways, the relevant dynamical characteristics and domains of structural stability of sociocultural systems. Social relations within and between even the most primitive human groups are always very complex in the sense that they involve large numbers of empirically significant variables at *all* levels of description. We try to cope with this situation by first considering biological mechanisms of the evolution of social behaviour from the perspectives of both adaptive change and structural differentiation. We then look for modes of cultural transformation of evolved social behaviour patterns. Eventually we employ these considerations to establish at least qualitative notions of structural stability and evolution applicable in sociocultural contexts.

Today behavioural ecology and sociobiology are largely based on concepts of evolutionarily stable strategies, or ESS's, of intraspecific competition and cooperation (Maynard Smith and Price 1973; Krebs and Davies 1978; Maynard Smith 1982). An ESS is a behavioural phenotype which, when fixed in a population, is maintained by natural selection against the spread of alternative behaviour patterns. More recently, Taylor and Jonker (1978), Zeeman (1980, 1981), Schuster et al. (1981) and Axelrod (1984) have developed the dynamical approach to evolutionary game theory ("evolutionary game dynamics"; Sect. IV.2) which fits into the general conceptual framework of biochemical, genetic and organismic population dynamics (Schuster and Sigmund 1983; Hofbauer and Sigmund 1984). The approach not only provides a complete description of the time evolution of behavioural phenotypes in populations exposed to natural selection, covering both pairwise and collective interactions (Maynard Smith 1982, pp. 23–27; Axelrod 1984). It also explains transitions between evolutionary games of different types under changing genetic and environmental constraints (Zeeman 1980, 1981; see also Sect. V.1). In fact, for an important class of evolutionary games frequently considered in the literature, the dynamics of a population game has been shown above to be structurally stable if the game admits an ESS (Sect. IV.2). Technically, the latter result implies that ESS theory can be used to characterise not only particular population dynamic systems of interacting behavioural phenotypes but also evolutionary transitions between such systems (non-ESS games may be structurally unstable). From a theoretical point of view, the result essentially means that species-specific behavioural phenotypes are invariant against, or may vary with, biomolecular evolution of the genome and changing environmental conditions, depending on whether these phenotypes are evolutionarily stable or not in the sense of game theory. For instance, preadptive behavioural traits are structurally stable since they persist under changing genetic and environmental constraints, adopting novel functions for which they have not been selected (cf. Wilson 1975, Chap. 3). The example of preadaptive functional change also shows that the conceptual framework of structural stability as applied to population dynamics encompasses phenomena of non-adaptive evolution as well. Recall that the concept of adaptation is used in the strict Darwinian sense here, that is, an organic trait is called adaptive only in relation to the function for which it has been designed by natural selection. This aspect, in turn, suggests that the concept of structural stability may serve as a useful analytical tool in evolutionary approaches to human social structures leading beyond the controversial issue of whether human behaviour in sociocultural contexts is adaptive in the sense of inclusive-fitness theory. In other words, the specific contribution to the theory of human social structure, which one may expect from sociobiology, may be found in explanations of the stability properties of evolved behaviour patterns.

Structural-stability approaches to human social organisation have previously been attempted at various degrees of elaboration (Deakin 1980; Weidlich and Haag 1983). A qualitative notion of the structural stability of sociocultural systems is appropriate to the purposes of the present investigation. Accordingly, we treat complex human society as a multicomponent system composed of individuals and small-scale subgroups. Characteristic constraints on this sytem, and on the interactions within it, are geographical circumscription, ecological pressures and exchanges with the sociocultural environment. We associate the microstates of society with the physiological and psychological motivational states (wishes, needs, thoughts and emotions) of social actors, and face-to-face relations between individuals. At the microlevel of organisation, perturbations of the social structure typically arise from individual actions that deviate from traditional patterns or violate the institutional norms of behaviour, and, at the macrolevel, from cultural innovations or environmental changes. It is then straightforward to call social relations and systems structurally stable or unstable in the technical sense in which the concept of structural stability has been introduced above.

The structural-stability/instability approach suggests itself to analyses of a broad range of phenomena of economic and sociocultural change. Recent applications to special cases include the stability of industrial power networks and of the world economy (Hastings 1984), stock exchange crashes and prison riots (for review and discussion of the relevant literature, see Deakin 1980). Here we use the approach to analyse a more abstract kind of social stability, namely the cross-cultural invariance and historical significance of types of social organisation. More precisely, invariance against changing cultural and historical conditions will be treated as an instance of the structural stability of social relations. We use the following comparative method of evaluating ranges and degrees of social stability in terms of differences in the effectiveness of alternative modes of social control. The manipulation of sociocultural relations generally operates on such various modes, involving institutions, norms (legal and others), traditions, authority, influence, power and individual incentives to join in economic organisations. Evidently, the effectiveness of any particular mode depends on organisational prerequisites, material means and specific circumstances, and so do the group structures involving this mode of social control. For example, gestural communication is a rather ineffective way to co-ordinate the behaviour of large numbers of individuals but proves highly effective in face-to-face interactions. Now one compares and classifies alternative procedures of social control according to their effectiveness and the sensitivity with which this effectiveness depends on means and circumstances. On the basis of this classification, one may then explain structural-stability properties of social systems. For instance, those modes of social behaviour manipulation, which are multifunctional and prove effective independently of particular sociocultural contexts, may be expected to govern historically and cross-culturally invariant types of social organisation. Conversely, phenomena of the geographical and cross-cultural stability of social patterns may become accessible to theoretical explanation when they can be shown to involve universally applicable means of social control.

Looking for biobehavioural stability factors of social patterns, we concentrate on ESS's of social interaction since these strategies show the necessary structural-invariance properties, as has been pointed out above. This procedure corresponds to the following view of human biosocial organisation. According to ethology and sociobiology, human social relations are networks of behaviour co-ordination and control operating on phylogenetically evolved response patterns with broad reaction ranges (Wilson 1975). Broadness of reaction ranges means a genetically constrained multiplicity of stimulus-response relations the functions of which include physiological regulation as well as institutional manipulations of human development. The phylogenetic heritage of the human species thus constitutes a set of preadaptive *instruments* of social organisation the functions of which, in cultural contexts, may or may not prove adaptive in the sense of Darwinian evolutionary theory, that is, these instruments may or may not be employed to maximise the reproductive fitness of social actors. Accordingly, phenomenological parallels between human and non-human behaviour need not always be understood as functional analogies through which culture simulates biological adaptations, as is frequently assumed in human sociobiology (Barash 1977, Chap. 10; Alexander 1979, p. 85; Irons 1979, p. 9). These parallels rather indicate that phylogenetic behavioural pre-adaptations may serve as *organic devices* suitable to engineer human relations for all conceivable, historically contingent purposes. General arguments and a striking example in support of this view of human sociobiology have also been worked out by Johnson (1986).

These conceptions and assumptions bear directly on the problems of the evolution and stability of complex hierarchical systems in cultural history. According to previous approaches to political anthropology (Flannery 1972), the rise of the state and similar sociopolitical organisations from the matrix of primitive egalitarian social relations can be explained as the culture-specific transformation of evolved face-to-face behaviours ("special-purpose patterns" in Flannery's terminology) into procedures of large-scale social control ("general-purpose patterns"). This transformation, in turn, will be argued to belong under the structural-instability type of evolution in the next section: The rise of civilisation was governed by non-directional ("random") forces, that is, historically and geographically contingent events like spontaneous cultural innovations and other unpredictable variations in the constraints on precivilised societies. Different levels of organisational complexity and stability arose from different modes of reorganising face-to-face patterns into effective means of more abstract social control. These means include the ritualistic reinforcement of the social order in communities intermediate between primitive and complex society, and the centralised, power-based governmental authority characteristic of states. Here we further elaborate on the biobehavioural factors which make political power a general-purpose mode of social control.

Eventually, the hypotheses and conclusions of the present section should be qualified in at least two respects. The assumption that early sociocultural evolution was driven by non-directional forces does not deny that human individuals and groups may show goal-directed behaviour and that the phenomena of social stability and change can be understood as effects of human purposes. However, social structures often remain reasonably stable despite historically and cross-culturally varying goals of social actors. Examples familiar to social anthropologists are the characteristics of political organisation on which the following subsections concentrate. In other words, it is largely a matter of conceptual topology and theoretical interest whether and to what extent human goals are made explicit. The structural-stability approach treats them as variables rather than fixed objects of concern.

A similar argument applies to the familiar distinction between structural and functional analysis, and the relevance of this distinction to the present way of conceptualisation. Social structures usually serve discernible functions, and functions (specific contributions of subunits to the global state and operation of the social system) can be viewed as interaction patterns (i.e., structures). So whether social relations appear as functions or structures depends on the context. Here we are concerned with interaction patterns that are multifunctional in the sense that they optimise social control under *varying* conditions.

2.3 Biobehavioural Bases of Influence and Power

For every social being, the range of his possible actions is at any time not only constrained by his physical environment but also by the other members of his group, their actions, wishes, needs, and so forth. These constraints are usually referred to as environmental and social *influences* respectively. If by choosing a certain action a person is able to determine the behaviour of others, the former is said to have *power* over the latter. *Coercion*, or *coercive power*, is said to be imposed on a person's actions if someone is able to determine them by whatever means, regardless of the person's consent or will (De Crespigny 1968; Pennock and Chapman 1972). Different social relations often depend on different types of power and influence to quite distinct degrees. In history one finds all kinds of domination imposed on social groups and their territories. Domination may be based on economic dependence of the ruled on the rulers, religious belief in the devine descent of the latter, custom or democratic consent. Some forms of domination are distinguished from others by the dominators and their administrations' readiness to resort to means of coercion and physical force to manipulate the social order. According to Max Weber, a group of persons is referred to as a *political organisation* if it is subject to political domination, that is, its existence and order is continuously safeguarded within a given territorial area by the application and threat of physical force on the part of the administrative staff (Weber 1972, Part I, Chap. 1, Sect. 17).

Two remarks may help to substantiate the last definition. Firstly, as has been admitted by Weber himself and is re-emphasised by any other attempt to frame political theory conceptually, the structure and function of political organisation cannot be explained only by specific modes of the enforcement of the social order. However, the emphasis Weber and others gave to power and physical force in defining the concept of politics admits a particularly clear, operational distinction of political corporate groups from other social organisations. For political communities show empirical properties with regard to the distribution and function of power within them and within the compass of their social environments which make them unique among human social groups. Secondly, by adopting the Weberian concept of political organisation we exclude from the analysis anthropologically highly significant types of social organisation such as rank society and the so-called segmentary and feudal societies (Flannery 1972; Adams 1975; Service 1975). But it is well known that these communal systems are distinct from complex society by size, degree of organisational differentiation and governmental structure. So it is not unreasonable to use a concept of political organisation that is sufficiently narrow so as to bring out this distinctness. That intermediate-scale, prepolitical community structures, too, often prove markedly stable without the use of centralised power is a different issue (see Sect. VI.3) and does not necessarily say anything about the structural stability of large-scale complex society.

Comparative ethological research has demonstrated that the relationships of influence, power and authority within human groups possess marked biological correlates in the social behaviour of other primate species. These correlates include aggression-submission patterns in dominance orders, the formation of coalitions, leadership and activities attracting social attention (Chance and Larsen 1976; Popp and Devore 1979; Omark et al. 1980; de Waal 1982; Barchas 1984); their occurrence and function in behaviour control often vary with ecological conditions and social contexts. The ethological notions of dominance ranking, dominance aggression and attention structure have been criticised repeatedly because of this variability and occasional lack of observational coherence in dominance rankings (for review and discussion, see Wittenberger 1981, pp. 591–594), and because of lack of logical rigour in certain expositions of attention structure theory (Schubert 1983). Concepts of social dominance and their correlates nonetheless seem to provide theories of intraspecific competition with an almost indispensable analytical framework. For a long time, the very problem of the evolutionary approach to social dominance was rather posed by the question of how to explain the selective advantages and adaptive value of ritualised submission patterns. For rank order in co-operative animal groups imposes fitness restraints on subordinate individuals who nonetheless appear in social-animal populations in more or less stable proportions from one generation to the next. A theoretically most satisfying and elegant solution of this problem is due to Maynard Smith and Price (1973) who have shown that genetically heritable behavioural differences in intraspecific competition for resources and reproductive success may give rise to ESS's. Unfortunately, in the social primates, especially in man, an embarrassing complexity and plasticity of social traits leave the concept of ESS hard to operationalise. However, since in non-human primates and in man dominance-submission responses tend to correlate with certain gross patterns of behaviour such as gestural displays, ritualistic status signalling and dominance aggression (Wilson 1975), which are frequently observed in other social species in similar contexts, ESS theory may still be of heuristic value in biobehavioural approaches to human dominance relations.

Evolutionary game theory and game dynamics have primarily been used and discussed in the literature from the perspectives of the natural selection of co-operative behaviour in groups of selfish individuals, with equally matched competitors ("symmetric games"). However, as first pointed out by Parker (1974) and Maynard Smith and Parker (1976), most social conflicts in real populations will occur between differentially matched opponents ("asymmetric games"). Circumstantial asymmetry in the chance of winning a contest, and differential fighting ability resulting from differences in age, sex, aggressiveness, body size or experience, can and usually will influence the resolution of animal dominance conflicts. Now the important point is that asymmetric population games tend to produce evolutionarily stable conflict strategies of the "common sense" type, "be prepared to escalate when the expected excess pay-off in fitness is positive, or else retreat". This tendency is due to the fact that asymmetric contest conditions may serve as "cues" by which dominance conflicts can be settled conventionally at minimum costs in fitness for the opponents. In human societies, an individual's capacity to succeed in social conflicts is co-determined, in culture-specific ways, by institutional systems, learning, traditions, and distributions of economic wealth, tools and weapons. Accordingly, cultural systems of cues, incentives and sanctions impose asymmetry conditions on social conflicts, that is, assign different ranges of social action to different individuals, thus generating certain gross modes of social control and conflict resolution which one usually associates with influence and power.

A few examples selected from classical accounts of human social organisation may illustrate this conclusion briefly. According to Weber (1972, esp. Part I, Chap. 3, Sect. 1), domination frequently involves some minimum degree of voluntary subordination on the part of those group members who are expected to obey. In such cases influence consists in persuasion, manipulation, inducement, and so forth. In Weber's typology of legitimate domination, charismatic authority is defined as the emotional stimulus to voluntary submission under the command of a person who is believed to possess extraordinary qualities. The pattern of charismatic leadership and submission fits into the pattern of ritualistic behaviour manipulation in animal groups, with conventional displays indicating an animal's superior physical and emotional dispositions to control the actions of the other group members (Dawkins and Krebs 1978). Remarkably enough, Weber (1972, Part I, Chap. 1, Sect. 1) himself foresaw the possibility that major features of charismatic authority (*breite Schichten des "Charisma"*) may admit biobehavioural explanations. Influence and power relations may also result from interest constellations, in particular, in economic contexts (Weber 1972, Part II, Chap. 8, Sect. 6). Since in animals and man the control of territories, tools and resources need not be correlated with expressions of specific organic traits, differential access to resources introduces an "uncorrelated asymmetry" into intraspecific conflicts which has been demonstrated to generate evolutionarily stable dominance-submission responses (Maynard Smith and Parker 1976). Examples of this type of conflict resolution have been reported from various, including primate, species (for review and discussion, see Dawkins and Krebs 1978; Maynard Smith 1982, Chap. 8). In human societies, economic wealth and resources are largely distributed according to unbalanced exchanges in the market so that the market situation becomes an important agent in the establishment of dominance relations, as is emphasised in Weber's theory of economic power. Viewed from biosocial perspectives, the market thus acts as a source of "uncorrelated asymmetries" imposing culture-specific constraints on evolved patterns of human dominance behaviour.

It is straightforward to extend these considerations to the patterns of coercive power in institutional systems in which co-operation is enforced either non-violently or violently. From a functional point of view, the institutionalised threat of sanctions in human social behaviour control is the cultural transformation of bully tactics and agonistic postures which act as ritualistic cues to settle conventionally animal and human face-to-face conflicts. This view of institutional repression is illustrated best in terms of an example by Deutsch (1963, p. 122) who refers to the familiar phenomenon that revolutions and insurgences can often be prevented by the mere display of those military means which would also be suitable to suppress them violently (patrolling army units in the riot-afflicted capital). As has frequently been noted, in man as a tool-using animal all conceivable objects in human environments can be easily turned into effective cues of appeasement when used in agonistic displays and threats of sanctions. The capacity of man to devise and use destructive weapons may indeed greatly reduce his dependence on biological disposition to succeed in conflict situations (Bandura 1979). But from the game-theoretical point of view, this reduced dependence is a general feature of animal contests involving uncorrelated asymmetries rather than a specific of human social conflict. Evolutionary game theory thus points to the biobehavioural origins of coercive power, and clarifies basic functional aspects of this mode of coercion in sociocultural contexts. The enforcement of co-operation in repressive institutions is a social control procedure operating on preadaptive human behavioural dispositions to form dominance orders; these dispositions are particularly easy to exploit in sociocultural systems because of the ease with which human individuals and groups can purposely engineer circumstances and conditions of social interactions so as to succeed in case of conflict. Secondly and more importantly, evolutionary game dynamics can show why conventional dominance-submission responses are virtually universal features of asymmetric intraspecific conflict, independently of whether the asymmetry is or is not correlated with the biological properties of interacting individuals (e.g., arises from culturally contingent attributes of social systems). As will be suggested in the next subsection, this independence is exactly what the structural stability of sociopolitical organisation draws upon.

2.4 Political Power

The historic rise of the state and civilised society from more primitive egalitarian and tribal community structures poses a multiplex stability problem. On the one hand, chiefdoms, pristine states and related types of ranked and stratified society are distinguished from less complex antecedent corporate groups by advanced degrees of division of labour, occupational specialisation of the group members and long-range social interactions well beyond kinship units (Flannery 1972; Service 1975). These relations depend sensitively on the reliability of the group members' future actions, which makes complex societies inherently dynamical systems (cf. Axelrod 1984, pp. 126–132, 174). Barter, trade and similar forms of the exchange of goods and services may give rise to highly interconnected networks of "balanced reciprocity", that is, transactions of mutually commensurate worth or utility among non-relatives. Even more complex modes of social reciprocation arise where interactions involve delays between giving and receiving goods or imply returns in terms of highly abstract obligations such as protection and justice. Complex society thus presents itself as a network of reciprocal expectations the establishment and consolidation of which poses a permanent problem.

On the other hand, one has the problem of what may be called the structural stability of political organisation. The cultural evolution of complex society was almost exclusively paralleled by the establishment of political domination, centralised government and formal law. These cross-culturally invariant features of sociopolitical evolution have stimulated extensive anthropological debates about whether there are universal primary causes, or "prime movers", of this process such as population growth, irrigation or warfare (Flannery 1972; Cohen and Service 1978; Claessen and

Skalník, 1978). However, as Cohen (1978) and others have argued, none of the demographic, cultural and ecological factors that have been considered in this context, provides a sufficient explanation of primary state formation; none appears to be uniformly correlated with the rise of all of the early states known to anthropologists, nor is it possible to decide in many well-documented cases whether these factors are preconditions or consequences of the establishment of centralised administration and political rule. Cohen (1978, p. 70) concludes that early state formation indeed has varied, geographically and historically contingent causes but tends to produce recurrent results, considering the striking similarity in the structure and organisation of primary states "so far removed from one another as Inca Peru, Ancient China, Egypt, Early Europe, or precolonial West Africa".

Now random structural variation producing stable, recurrent effects is the virtually universal evolutionary pattern of hierarchical differentiation to which structural-stability analyses specifically apply. With respect to the evoluton of complex society, such an analysis will have to explain the particular mode of social control on which governmental authority and political domination operate and to which they may owe their cross-cultural stability properties. Recall that the notion of the structural stability of political organisation does not mean here that large-scale political corporate groups or even whole civilisations possess an intrinsic capacity to resist historical change or decay. Nor does it suggest that political organisation belongs to the cultural universals of human society. It simply refers to the *empirical* fact that political organisation is a universal cultural feature of all *complex* societies, and offers a useful way of looking at this fact from an evolutionary point of view.

The anthropological data on state origin suggest that beyond any particular, historically realised mode of establishing and legitimising social order, political power (the capacity to enforce this order violently) confers the observed stability properties on large-scale society. From a comparative, functional point of view, this stabilising effect mainly depends on two factors; first, the means of the violent enforcement of social behaviour are highly impersonal, that is, apply to everybody to more or less the same degree, and secondly, they are suitable for almost all conceivable manipulative purposes. To use Deutsch's (1963, pp. 122-124) striking analogy, any means of physical force may act on people like gold or money: it is "acknowledged" by everybody and outweighs most individual values and interests. Alternative factors and functional bases of social control such as charismatic authority, economic interests and the legitimacy of institutions largely depend on subjective attitudes, needs and beliefs, and may thus vary with the individual motives of social action. This problem also bears on social relations which are formally constituted (e.g., legal forms of domination) and yet, which have to be actually maintained. As first pointed out most emphatically by Hobbes (*Leviathan*, Part II, Chap. 17) the validity of the legal order is ultimately not a matter of the normative content, legitimacy or rationality of the law but of effective measures to enforce this law in case of non-conformity motivated by individual self-interests. Basically, it is an immediate corollary to Hobbes' political anthropology that large-scale institutions are more effectively controlled the better they function independently of individual attitudes, interests and values.

The preceding considerations suggest that the empirical significance of physical force and institutional repression for the control of human social relations varies in proportion to the size and complexity of the society rather than with particular environmental conditions and historical circumstances. Once more we relate to the example of early state formation to provide evidence in support of this hypothesis. The increasing importance of routine, impersonal means and modes of social control, which characterises the route from ranked to fully stratified society, corresponds to a depersonalisation of social relationships at almost all levels of organisation. In chiefdoms and tribal societies with hereditary status order, domination is typically exercised by means of non-violent coercion and theocratic rule (the administration is largely identical with the priesthood). In pre-state societies and in primitive states

"the pressure on individuals to conform is a property of the local kin group... The considerable conformity is to the norms of the traditional folk society... The state rules through intermediaries, and the face-to-face 'despotism' is simply that of the communal kin group under the aegis of its local priestchief" (Service 1975, p. 301).

Now the first empirically relevant fact is that the rise of increasingly stratified political communities is paralleled by a tremendous growth in group size (Flannery 1972), which makes social interactions between non-kin more likely. Secondly, the establishment of abstract legal order as opposed to non-coercive custom, religious obligation, and ritually reinforced social norms generates an entirely novel regime of impersonal social relations. From his empirical surveys Service (1975, p. 90) concludes

"that the origins of the state may be accompanied by a sudden increase in the number of repressive laws, by more severe repression, and perhaps by new kinds of laws. And it is very likely that the new state will have a more visible, more formal and more explicit judicial and punitive machinery."

Thirdly, the strongly centralised administrations of large stratified and differentiated societies involve high degrees of depersonalised authority (authority attached to offices rather than officeholders). Finally, the development of a professional ruling elite largely divorced from the bonds of kinship, and of residential patterns based on occupational specialisation rather than blood or affinal ties (Flannery 1972), gives rise to one more fabric of impersonal social relations the formal constitution of which is almost universally protected by political domination.

According to Weber (1958, Chap. 4, Sect. 8; 1977), the evolution of the modern state since the European age of religious civil wars is characterised by an increasingly rational bureaucratic and economic organisation, and, more specifically, by the establishment of the monopoly and exclusive right of use of physical force on the part of the state administration. Viewed from the biobehavioural perspectives of the previous section, the state monopoly of physical force evolved so as to rationalise (i.e., increase the effectiveness of) political power, too. Political power is an impersonal multifunctional means to transform social conflicts into asymmetric games with conventional dominance-submission strategies as structurally stable solutions. Accordingly, monopolising the available means of coercive force is a social control strategy designed to maximise the degrees of asymmetry which power can impose on conflicts within and between political communities. Needless to say, this power-maximising strategy cannot be optimal and structurally stable in an absolute sense. It may, for instance, lead to arms races that escalate conflicts rather than settle them conventionally. Such an escalation will continue if during an arms race differences in disposition to win the contest are averaged out, political conflicts thus getting symmetrised. However, this effect merely reconfirms the present thesis that the splitting of political power or the failure to monopolise it tends to destabilise structurally not only asymmetric social games and their dynamics but also civilised society as such (cf. Hobbes, *Leviathan*, Part II, Chap. 29).

2.5 Summary and Discussion

Human face-to-face interaction and communication are less constitutive of social relations the larger the number of interacting individuals. Political power gains in importance as a means of social control the less human group structures depend on face-to-face interactions. The biobehavioural bases of political power analysed above account for the cross-culturally invariant function of violent coercion as an institutional sanction, and they explain the co-evolution of complex society and political organisation in cultural history. These conclusions are empirically supported by the presently available data on state origin, and by organisational optimality features of the modern Western State.

It is instructive to contrast the present analysis of the evolutionary origins and function of political power with previous biological approaches to sociopolitical organisation drawing on concepts of competitive selection and kin selection (Carneiro 1970, 1978; Alexander 1979; White 1981; Masters 1983). These approaches are based on the common hypothesis that size and complexity of co-operative groups are social adaptations, that is, shaped by the natural selection of human social behaviour and positively correlated with the survival and reproduction of kinship units and genealogical lineages within society. As noted above, the problem with these approaches is that hierarchical differentiation is generally not an adaptation process. A case in point is the cultural evolution of large-scale society involving political domination. For political domination can be, and has always been, used to establish and enforce relationships of both balanced and highly unbalanced reciprocation (Willhoite 1981; see also next section), its net effect on the inclusive fitness of social actors thus being far from indiscriminately beneficial (for an extreme standpoint on this issue, see van den Berghe 1981, Chap. 1, Sect. 4). Therefore, empirically there seems to be little point in viewing the state as an institution specifically designed to engender mutual benefits for its citizen members. Moreover, as Weber (1972, Part I, Chap. 1, Sect. 17) has emphasised, politics is an inherently multifunctional process and cannot be associated in a one-to-one fashion with any particular kind of purpose political corporate groups may serve.

The notion of the state as a large-scale social system designed to benefit inclusive fitness rather insinuates a regression to the philosophy of natural right and the Aristotelian tradition of normative political theory, with individual welfare as a biologically circumscribed measure of what is good and just in politics (e.g., Masters 1981, 1983). To be sure, none of the classical political writers who – like Machiavelli, Hobbes and Weber – do not fit into this tradition ever denied that by co-ordinating social behaviour the state may create collective goods and mutual benefits for all citizens. But these authors also insist on the point that the nature and function of politics must be searched for beyond good and evil: It is the *empirical* fact of the broadness of human behavioural response ranges, that is, the "plasticity of human nature" which, from despotism to the Welfare State, leaves any form of complex society unstable that does not include an effective power structure (Machiavelli,

Discourses, Part I, Chap. 55; Part II, Introduction; *The Prince*, Chap. 6; Hobbes, *Leviathan*, Part II, Chaps. 17, 29).

3 On the Evolution of Complex, Political Society

Today archeologists and ethnologists widely accept that the establishment of multicommunity chiefdoms with hereditary rank order, theocratic rule and certain primitive patterns of centralised administration preceded virtually all of the primary states known in history (Flannery 1972; Service 1975). Acceptance of this position has simplified the investigation of the origin of the state considerably. Since the state is a territorial organisation with a highly centralised government, a professional ruling class and coercive law, investigation now concentrates on the transformation of chiefly authority and theocratic domination into political power as an institutionalised sanction (Service 1975).

However, as pointed out in the preceding section, the detailed processes which brought about this transformation still remain a matter of controversy among anthropologists. Present-day theories of the origin of the state can be classified into those that imply sociocultural relations and processes exclusively, and those that invoke evolutionary biological categories (e.g., Carneiro 1970, 1978; Alexander 1979). Other hypotheses attempt to account for primary state formation in terms of "prime movers" such as irrigation, warfare or population growth (for a review, see Flannery 1972) as opposed to systemic, multivariate approaches which exclude monocausal explanations (Flannery 1972; Webster 1975; Wright 1977).

Persuing the argument of the preceding section, we suggest that the rise of stratified, political society in cultural history fits into the general evolutionary scheme of structurally unstable systems in the prebiotic and living world. Once more, part of the conceptual framework applied in the analysis is borrowed from dynamical-systems theory (biophysics, evolutionary ecology) and is reinterpreted with reference to human social interactions. We shall argue that the evolutionary approach to primary state formation is consistent with the "systemic" position in political anthropology. This position views the origin of the state as a result of multiplex interactions between social, cultural and ecological factors and variables. Complex political society has gained significance in cultural history by developing strategies of social cohesiveness and stability suited to cope with environmental uncertainty and change.

3.1 The Evolutionary Conceptual Framework

Following anthropologist Kent Flannery, we have argued above that the rise of the state and similar forms of complex social organisation can be understood as the transformation of the special-purpose patterns of face-to-face interaction into general-purpose patterns of large-scale social control. This transformation, which has been called "promotion" by Flannery (1972), can be usefully applied in biobehavioural analyses in the following way: According to sociobiology theory, there is hereditary variation in human social traits, leaving these traits adaptive in primitive societies exposed to strong ecological pressure. However, human social reaction ranges are generally broad, that is, multifunctional. Human response patterns such as kin altruism, ritualised communication and dominance-submission interactions may thus serve functions for which they have not been designed by natural selection. In other words, they may control adaptive and non-adaptive, primitive and complex social relations. Promotion then includes preadaptive functional changes in human social behaviour and social control, i.e., the transformation of adaptive behaviours into non-adaptive ones in sociocultural contexts.

The concept of promotion is significant to political anthropology for at least two reasons. Firstly, it provides anthropology theory with an evolutionary framework to describe the increasing hierarchical complexity in sociocultural systems. Like any other structure in the living world, social structures are exposed to environmental changes which, from many research perspectives, are reasonably assumed to be stochastic. Differential stability and survival of newly arising systems then depend on the capacity of each particular system to reorganise existing ("preadaptive") modes of social behaviour control. The trial-and-error type of evolution accordingly governs not only biophysical processes but also increasingly complex hierarchical organisation in cultural history. In particular, a comparative approach to the stability of large-scale complex social systems has been suggested above and has been used to evaluate the relative success of evolutionary steps in sociohistorical contexts. The approach relates to the biobehavioural preadaptations of culture and, as has been attempted to show, helps to explain the coincidence of the historic rise of large-scale social groups, on the one hand, and political power as an institutionalised sanction, on the other.

The second advantage which the concept of promotion, or functional shift, offers to theoretical anthropology is this concept's compatibility with *stochastic* evolutionary theories. As has been emphasised above, the same point can be made regarding the concepts of structural stability and instability of dynamical systems. Roughly speaking, by stochastic (indeterministic, probabilistic) theories one means theories whose empirical postulates are not (deterministic) cause-effect implications. This type of theory suggests itself to the social sciences and to historical disciplines because historical events tend to be highly singular and often cannot be attributed to identifiable causes in definite ways. Theories of promotion in evolutionary anthropology can avoid these partly logical and partly observational difficulties. The changes in social patterns and functions with which theories of promotion are concerned may be intrinsically random. What these theories can do, then, is to assign different attributes and stability properties to different outcomes of historical events and processes, however improbable each particular event may be. According to the methodological standards of empirical science, nothing is required here except that the assignments are hypotheses testable by observation.

As for sociocultural evolution, we do not attempt to offer deterministic theories here. The present argument rather runs as follows. Whatever the spatiotemporally contingent "causes" of the formation of large-scale social groups may be, those *forms* of complex social organisation will prove the historically most significant ones that are usually subsumed under the concepts of political organisation and state. Logically, this hypothesis is based on considerations of the stability of social structures: Institutionalised (or "legal") authority is easier to maintain in large groups than charismatic authority (Max Weber's argument) although the former is the historical, cultural transformation of the latter. In a similar fashion, we compare codified with ritualised social norms, political power with face-to-face dominance conflict, and so forth. Empirically, the hypothesis can be tested against the anthropological and historical data.

3.2 Biobehavioural Bases of Sociocultural Complexity

In the "systemic" approach to political anthropology, theoretical interest in the rise of the state concentrates on the ways in which control processes operate in increasingly complex societies. Primitive egalitarian societies (hunter-gatherer bands, horticulturalists and primitive agriculturalists) typically comprise autonomous groups formed through the familial bonds of kinship and marriage with weakly developed memories of descent and ancestral lineage. Intragroup conflict as well as intergroup aggression are by no means as rare as some anthropologists currently maintain (Ember 1978). Division of labour is based on age and sex. Moderately elaborate ritual is employed as a means of social communication. Evidence from sociobiology suggests that basic social traits of egalitarian societies tend to involve local adaptations of species-specific patterns of co-operation, competition and communication.

The transition from egalitarian society to rank societies and chiefdoms is marked by a conspicuous increase in average population size and internal differentiation of society with regard to division of labour and administrative functions. Leadership, economic wealth and superior social rank become attributed to noble lineages. As Flannery (1972) notes, the origins of hereditary inequality are difficult to explain. However, the elements of hereditary status arrangements – dominance orders and the constraints of kinship and descent – are not unique to rank society. Comparative ethology and sociobiology rather suggest that they constitute much of homologous and analogous primate social structure. Within egalitarian kinship units, where the degrees of relatedness vary less significantly over the group, dominance orders and division of labour may therefore be expected to be constrained predominantly by the preadaptations of the individual members' age and sex. Moreover, in multifamily groups the allocation of social rank and resources to whole families rather than individuals may well constitute adaptations that are relevant to an analysis based on inclusive-fitness theory. Alexander's (1979) account of nepotism (dispensing benefits to relatives other than immediate descendants) is largely concerned with this kind of allocation. On the other hand, it is paradoxical that biosocial adaptations should explain the evolution of social structures which, because they involve marked, rank-specific differences in access to goods and resources, impose conspicuously counteradaptive constraints on the survival and reproductive efforts of lower-rank individuals. To be sure, inclusive-fitness theory explains why group life can be beneficial even to subordinate individuals and subgroups. But this theory hardly explains why, by virtue of cultural inheritance of social status, group life should be *more* advantageous for one subgroup than for most others. This situation is correctly described by Willhoite's (1981) "dilemmas of rank and reciprocity".

Now the present structural-stability approach to social differentiation starts from the fact that hereditary inequality is one of the most prominent features in the history of civilisation. This empirical fact is then interpreted as an effect of preadaptive dispositions in human social behaviour patterns (dominance, submission and the awareness of kinship bonds). In rank society, these dispositions have been effectively extended, in culture-specific ways, beyond the ranges of face-to-face relations to more inclusive social units. This extension is clearly an instance of promotion in Flannery's (1972) sense, namely the rise of an institution from one level of social control to another, superior level. In particular, Flannery refers to the establishment of the "general-purpose" offices of chieftainship and sanctified noble lineages from the "special-purpose" face-to-face leadership of the informal headman in a primitive hunter-gatherer society.

As has been discussed in the previous section, similar modes of sociocultural evolution govern the transition from chiefdoms into primary states. Comparison of social rank ordering and stratification in prestate societies and primary states shows that the social response patterns entering political domination are not specific of political organisation. These responses clearly have roots and parallels in theocratic rule and authority in chiefdoms. In particular, domination in prestate societies is typically reinforced through elaborate rituals and religious practices (Flannery 1972; Service 1975), while in political communities sympathetic ritual as a stimulus of coherent social action has become transformed into abstract legal norms. The chief's charismatic or traditional authority is allocated to state offices rather than officeholders. Authority thus becomes depersonalised, as Max Weber emphasised in his classic account of legitimate domination.

3.3 From Rites to Sanctions

Ritualisation, that is, the evolution of previously non-communicative traits into species-specific signals, is an important mode of the Darwinian evolution of social communication (Wilson 1975; Dawkins and Krebs 1978). Certain general aspects of the ritualisation of behaviour patterns are useful in understanding human sociocultural development (Wilson 1975, p. 244; Eibl-Eibesfeldt 1979, 1984, Chap. 6). Ritualisation implies functional shifts in preadaptive traits similar to the modes of sociocultural change (promotion) that are to be observed in the rise of complex society. In fact, in non-human species ritualisation often leads to elaborate interaction patterns within given response ranges constrained by the genome. Furthermore, in ritualisation emotional movements are turned into symbols (signals) whose meanings vary with the contexts in which they are displayed in animal groups. Finally, perception and identification of species-specific stimuli and signals depend on the detector devices and motivational states of the psychophysiological apparatus. As Wilson (1975, p. 224) notes, in animal communication the effects of ritualisation are much more frequently to be observed than the effects of the alternative process, namely the adaptation of sensory organs and physiological releasing mechanisms to pre-existing communicative displays.

Phylogenetic ritualisation thus provides an example for the present hypothesis that evolutionary invariant patterns result where differential rates of change arise between evolving communicative and manipulative traits with high individual flexibility, on the one hand, and their genetic and physiological constraints, or biological "inertia", on the other. This mode of differential structural change must be expected to govern also human biosocial evolution, which involves both biological "inertia" and disproportionally broad individual response ranges. Therefore, it is useful to examine the role of the invariant biobehavioural factors governing the development of legal systems and institutions from the ritual practices and ceremonies which in prestate societies serve to stabilise the social structure, control dominance orders and shape consent within the group (Burns and Laughlin 1979). Because there exist basic psychological correlates of ritualistic behaviour, comparative ethologists and social psychologists tend to assume that ritual rests on complex but discernible biological substrata. In particular, neuroanatomical and neurophysiological approaches to human ritualistic behaviour seem to emphasise the sensitising impact of external (social and physiological) stressors on the nervous system and the constitutive functions of each hemisphere of the human brain in generating and screening ritualistic cues, movements and cognitive processes (d'Aquili 1978; Lex 1978, 1979). Moreover, ethological and sociobiological evidence suggests that expressive movements and symbols in human ritual act so as to stimulate phylogenetically analogous and homologous releasers that in animals and man correlate with the patterns of status signalling, aggressive displays, facial expressions, intentional movements and play behaviour (Wilson 1975; Dawkins and Krebs 1978; Eibl-Eibesfeldt 1979, 1984; Fagen 1981).

In the cultural evolution of complex, political society, the communicative and manipulative functions of human ritual are thus each conferred on a different type of institution. The basic norms of the society become largely coded as laws, while the legal order tends to be reaffirmed by the coercive institutions of the state. Again the process involves rearrangements of pre-existing special-purpose patterns to more general modes of social control (Service 1975, Chap. 4). Discussing legal versus non-legal reinforcement, Service stresses the increasing significance of explicit definitions of the social norms. Sanctified custom, that is, ritualistic interaction patterns which have been supranaturally sanctioned, become reduced to abstract, codified, formal law. Formal and codified modes of social regulation are particularly suitable for controlling groups too large and too diversified to communicate effectively their basic cultural norms and values by means of ritualistic displays.

3.4 Summary

Human ritual and its cultural transformations have been interpreted as cross-culturally invariant, biobehavioural concomitants of complex, political society. As in the previous section, where political power has been considered as an evolutionary structure, we have argued that under varying historical and environmental conditions sociocultural structures are more stable the more effectively they exploit human species-specific behavioural dispositions by turning them into instruments of large-scale social control. In fact, empirical data from political anthropology and comparative ethology and sociobiology suggest that the distinction between ranked and fully stratified society corresponds to distinct stability factors at different levels of sociocultural complexity. The observable gross modes of sociopolitical control operate on biobehavioural dispositions in level-specific ways.

VII Concluding Remarks

The preceding evolutionary analyses have been carried out from various logical and theoretical perspectives on increasing biosocial organisation in natural and cultural history. As for the material problems involved, a type of unified theory has been developed which conveys system – theoretical aspects of natural self-organisation, but which proved applicable to sociocultural evolution as well.

The philosophical part of the inquiry has been concerned with concepts of emergent evolution, the holism-reductionism controversy in the ongoing sociobiology debate, and with reconstructions of these issues in exact logical and semantical terms. It has been shown that the basic principles of holism and emergence are trivial or false if not non-sensical. At any rate, the concept of emergence has been proved inherently incompatible with empirical, scientific evolutionism. Although we did not take a "reductionist" position here, our unified theories of various aspects of biosocial evolution did not plainly exclude the "reductionist" and "selectionist" theories of sociobiology. In fact, all that we did was to consider parameter families of selection equations (Chaps. IV, V) rather than a particular kind of selection equation. This simple shift in theoretical perspective endowed us with a unifying approach to a number of problems of increasing hierarchical organisation, while retaining the entire powerful apparatus of neo-Darwinian adaptation theory.

The present type of unified theory implies conceptual idealisations and approximate assumptions demanding an overall evaluation of the results of Part Two. The parametrised laws and systems introduced above have proved particularly suitable for describing evolutionary processes in non-dynamical ("secular") time scales. However, when little or nothing is known about the actual parameter values and parameter changes of real organic systems in natural history, the macroevolution of these systems could not be predicted (or retrodicted) even if their dynamics were well understood. The present type of parametrised laws may then be useful at least in theoretical analyses of the possible modes of long-term evolution, as the examples in Section V.2 demonstrate. On the other hand, if simple heuristic assumptions about control parameters can be made (e.g., $\mu =$ constant in (V. 82)), a basic understanding of macroevolutionary mechanisms such as autocatalysis in hominid cultural ecology may become possible.

Stochastic evolutionary processes and, with the exception of the first example in Section II.1.1, discrete dynamical systems have not been considered, although we have admitted that both are empirically significant for natural self-organisation. However, the effects of adaptation and structural instability, which we have been given a unifying theoretical treatment here, may become even more pronounced in discrete and in stochastic systems. For instance, Templeton (1982) has argued that genetic drift as a random process may well act so as to accelerate natural selection and adaptation rather than to interfere with them. And it is well known that

Concluding Remarks

discrete-time dynamical systems are generally much more structurally unstable than their continuous counterparts (May 1976). Our conclusions may therefore be expected to be broader in range and validity than the particular ways we obtained them seem to suggest.

References

Adams EW (1959) The foundation of rigid body mechanics and the derivation of its laws from those of particle mechanics. In: Henkin L, Suppes P, Tarski A (eds) The axiomatic method. North Holland, Amst, pp 250–265
Adams RN (1975) Energy and structure: a theory of social power. Texas Univ Press, Austin
Akin E (1979) The geometry of population genetics. Springer, Berlin Heidelberg New York
Alexander RD (1979) Darwinism and human affairs. Univ Washington Press, Seattle
Allen PM (1976) Evolution, population dynamics, and stability. Proc Natl Acad Sci USA 73: 665–668
Axelrod R (1984) The evolution of cooperation. Basic Books, NY
Babloyantz A (1986) Molecules, dynamics, and life. Wiley, NY
Balzer W, Pearce DA, Schmidt H-J (eds) (1984) Reduction in science. Reidel, Dordrecht
Bandura A (1979) Psychological mechanisms of aggression. In: Von Cranach M, Foppa K, Lepenies W, Ploog D (eds) Human ethology, Cambridge Univ Press, Cambridge, pp 316–379
Barash DP (1977) Sociobiology and behavior. Elsevier, NY
Barchas PR (ed) (1984) Social hierarchies. Greenwood Press, Westport, Conn
Barkow JH (1980) Sociobiology: is this the new theory of human nature? In: Montagu A (ed) Sociobiology examined, Oxford Univ Press, Oxford, pp 171–197
Bechtel W (ed) (1986) Integrating scientific disciplines. Nijhoff, Dordrecht
Bomze I, Schuster P, Sigmund K (1983) The role of Mendelian genetics in strategic models on animal behaviour. J Theor Biol 101: 19–38
Bonner JT (1980) The evolution of culture in animals. Princeton Univ Press, Princeton
Bonner JT (1988) The evolution of complexity by means of natural selection. Princeton Univ Press, Princeton
Boorman SA, Levitt PR (1980) The genetics of altruism. Academic Press, NY
Bourke AFG (1988) Worker reproduction in the higher eusocial Hymenoptera. Q Rev Biol 63: 291–311
Bowler PJ (1984) Evolution: the history of an idea. Univ California Press, Berkeley
Boyd R, Richerson PJ (1985) Culture and the evolutionary process. Chicago Univ Press, Chicago
Brian MV (1979) Caste differentiation and division of labor. In: Hermann HR (ed) Social insects Vol. I. Academic Press, Lond NY pp 121–222
Brown RLW (1983) Evolutionary game dynamics in diploid populations. Theor Popul Biol 24: 313–322
Burns T, Laughlin CD (1979) Ritual and social power. In: d'Aquili EG, Laughlin CD, McManus J (eds) The spectrum of ritual: a biogenetic structural analysis. Columbia Univ Press, NY, pp 249–279
Campbell DT (1974) Downward causation. In: Dobshansky T, Ayala FJ (eds) Hierarchically organized biological systems. Macmillan, Lond, pp 179–186
Caplan AL (ed) (1978) The sociobiology debate. Harper and Row NY Hagerstown Lond
Carneiro RL (1970) A theory of the origin of the state. Science 169: 733–738
Carneiro RL (1978) Political expansion as an expression of the principle of competitive exclusion. In: Cohen R, Service ER (eds) Origin of the state. Inst Stud Hum Issues, Philadelphia, pp 205–223
Carson HL (1975) The genetics of speciation at the diploid level. Am Nat 109: 83–92
Causey RL (1977) Unity of science. Reidel, Dordrecht
Cavalli-Sforza LL, Feldman MW (1981) Cultural transmission and evolution. Princeton Univ Press, Princeton

Chagnon NA, Irons W (eds) (1979) Evolutionary biology and human social behavior. Duxbury Press, North Scituate, Mass
Chance MRA, Larsen RR (eds) (1976) The social structure of attention. Wiley, NY
Charlesworth B (1980) Models of kin selection. In: Markl H (ed) Evolution of social behavior. Chemie, Weinheim, pp 11–26
Claessen HJM, Skalnik P (eds) (1978) The early states. Mouton, The Hague
Cloninger CR, Yokoyama S (1981) The channeling of social behavior. Science 213: 749–751
Cohen MN (1980) Speculations on the evolution of density measurement and population regulation in *Homo sapiens*. In: Cohen MN, Malpass RS, Klein HG (eds) Biosocial mechanisms of population regulation. Yale Univ Press, New Haven, pp 275–303
Cohen MN (1981) Comments. Curr Anthropol 22: 532
Cohen MN, Malpass RS, Klein HG (eds) (1980) Biosocial mechanisms of population regulation. Yale Univ Press, New Haven
Cohen R (1978) State origins: a reappraisal. In: Claessen HJM, Skalnik P (eds) The early states Mouton, The Hague, pp 31–75
Cohen R, Service ER (eds) (1978) Origins of the state. Inst Stud Hum Issues, Philadelphia
Corning PA (1983) The synergism hypothesis. McGraw-Hill, NY
Cunningham WJ (1955) Simultaneous non-linear equations of growth. Bull Math Biophys 17: 101–110
Czaplewski RL (1973) A methodology for evaluation of parent-mutant competition using a generalized non-linear ecosystem model. J. Theor Biol 40: 429–439
d'Aquili EG (1978) The neurobiological bases of myth and concepts of deity. Zygon 13: 257–275
Darwin C (1859) The origin of species by means of natural selection. Murray, Lond
Dawkins R (1976) The selfish gene. Oxford Univ Press, Oxford
Dawkins R (1982) The extended phenotype. Freeman, Oxford
Dawkins R, Krebs JR (1978) Animal signals: information or manipulation? In: Krebs JR, Davies NB (eds) Behavioural ecology. Blackwell, Oxford, pp 282–309
Deakin MAB (1980) Applied catastrophe theory in the social and biological sciences. Bull Math Biol 42: 647–679
De Crespigny A (1968) Power and its forms. Political Stud 16: 192–205
Deutsch KW (1963) The nerves of government. Free Press, NY
de Waal FBM (1982) Chimpanzee politics: power and sex among apes. Cape, Lond
Dodson MM (1976) Darwin's law of natural selection and Thom's theory of catastrophe. Math Biosci 28: 234–274
Druss M (1981) Comments. Curr Anthropol 22: 532–533
Ebbinghaus H-D, Flum J, Thomas W (1984) Mathematical logic. Springer, Berlin Heidelberg New York Tokyo
Eibl-Eibesfeldt I (1979) Ritual and ritualization from a biological perspective. In: von Cranach M, Foppa K, Lepenies W, Ploog D (eds) Human ethology. Cambridge Univ Press, Cambridge, pp 3–55
Eibl-Eibesfeldt I (1980) Grundriß der vergleichenden Verhaltensforschung 6 Auf. Piper München
Eibl-Eibesfeldt I (1984) Die Biologie des menschlichen Verhaltens. Piper München. (transl) (1989) Human ethology. De Gruyter, Hawthorne
Eigen M (1971) Self-organization of matter and the evolution of biological macromolecules. Naturwissenschaften 58: 465–522
Eigen M (1983) Self-replication and molecular evolution. In: Bendall DS (ed) Evolution from molecules to men. Cambridge Univ Press, Cambridge, pp 105–130
Eigen M, Schuster P (1979) The hypercycle. Springer, Berlin Heidelberg New York
Eigen M, Winkler R (1975) Das Spiel. Piper München. (transl) (1981) Laws of the game. Knopf, NY
Eldredge N, Gould SJ (1972) Punctuated equilibria: an alternative to phyletic gradualism. In: Schopf TJM (ed) Models in paleobiology. Freeman, San Francisco
Elredge N, Tattersall I (1982) The myths of human evolution. Columbia Univ Press, NY
Elsasser WM (1975) The chief abstractions of biology. North Holland, Amst
Ember CR (1978) Myths about hunter-gatherers. Ethnology 17: 439–448
Eshel I (1982) Evolutionarily stable strategies and viability selection in Mendelian populations. Theor Popul Biol 22: 204–217

Essock-Vitale SM, McGuire MT (1980) Predictions derived from the theories of kin selection and reciprocation assessed by anthropological data. Ethol Sociobiol 1: 233–243
Ewens WJ (1979) Math popul genet. Springer, Berlin Heidelberg New York
Fagen R (1981) Animal play behavior. Oxford Univ Press, NY
Findlay CS, Hansell RIC, Lumsden CJ (1989) Behavioural evolution and biocultural games: oblique and horizontal cultural transmission. J Theor Biol 137: 245–269
Flannery KV (1972) The cultural evolution of civilizations. Annu Rev Ecol Syst. 3: 399–426
Fox R (1979) Kinship categories as natural categories. In: Chagnon NA, Irons W (eds) Evolutionary biology and human social behavior. Duxbury Press, North Scituate, Mass, pp 132–144
Fried MH (1967) The evolution of political society. Random House, NY
Geiger G (1983) On the dynamics of evolutionary discontinuities. Math Biosci 67: 59–79
Geiger G (1985a) Autocatalysis in cultural ecology. BioSystems 17: 259–272
Geiger G (1985b) The concept of evolution and early state formation. Politics life sci 3: 163–181 (including two commentaries and author's response)
Geiger G (1985c) Is life a zero sum game? (book review). Politics life sci 4: 80–81
Geiger G (1986) Der Begriff der einheitlichen Theorie – eine modelltheoretische Untersuchung zur Struktur und Evolution hierarchischer Systeme. Habilitationsschrift, Techn Univ München
Geiger G (1988a) Synthesis of theories through parametrisation of laws. I. Basic Definitions. Erkenntnis 29: 343–355
Geiger G (1988b) Synthesis of theories through parametrisation of laws. II. Example: Neo-Darwinian synthetic theory. Erkenntnis 29: 357–368
Geiger G (1988c) On the evolutionary origins and function of political power. J Soc Biol Struct 11: 235–250
Geiger G (1989) Sociobiology and the structural stability of behavior patterns. Math Biosci 93: 117–145
Ginzburg LR (1977) The equilibrium and stability for alleles under density-dependent selection. J Theor Biol 68: 545–550
Ginzburg LR (1983) Theory of natural selection and population growth. Benjamin-Cummings, Menlo Park, Ca
Goel NS, Maitra SC, Montroll EW (1971) On the Volterra and other Nonlinear models of interaction. Rev Mod Phys 43: 231–276
Gould SJ (1977) The return of hopeful monsters. Nat Hist 86: 22–30
Gould SJ (1980) Is a new and general theory of evolution emerging? Paleobiology 6: 119–130
Gould SJ (1982) Darwinism and the expansion of evolutionary theory. Science 216: 380–387
Gould SJ, Eldredge N (1977) Punctuated equilibria: tempo and mode of evolution reconsidered. Paleobiology 3: 115–151
Gould SJ, Lewontin RC (1979) The spandrels of San Marco and the Panglossian paradigm: a critique of the adaptationist programme. Proc R Soc Lond 205: 581–598
Gowlett J (1984) Ascent to civilization. Knopf, NY
Grafen A (1985a) A geometric view of relatedness. In: Dawkins R, Ridley M (eds) Oxford surveys in evolutionary biology, Vol 2. Oxford Univ Press, Oxford, pp. 28–89
Grafen A (1985b) Hamilton's rule o.k. Nature (Lond) 318: 310–311
Greenwood DJ Stini WA (1977) Nature, culture, and human history. Harper and Row, NY Hagerstown Lond
Grotemeyer K-P (1969) Topologie. Bibliogr Inst, Mannheim
Gurel O, Rössler OE (eds) (1979) Bifurcation theory and applications in scientific disciplines. NY Acad Sci, NY
Haken H (1978) Synergetics, 2nd edn. Springer, Berlin Heidelberg New York
Haken H (1981) Erfolgsgeheimnisse der Natur. Dtsch Verlags-Anst, Stuttgart
Haken H (1983) Advanced synergetics. Springer Berlin Heidelberg New York
Hamilton WD (1964) The genetical evolution of social behaviour I, II. J Theory Biol 7: 1–52
Harris M (1979) Cultural materialism: the struggle for a science of culture. Random House, NY
Hart H (1959) Social theory and social change, In: Gross L (ed) Symposium on social theory. Row, Peterson and Co, NY, pp 196–283
Hassan FA (1980) The growth and regulation of human population in prehistoric time. In: Cohen

MN, Malpass RS, Klein HG (eds) Biosocial mechanisms of population regulation. Yale Univ Press, New Haven, pp 305–319

Hastings HM (1984) Stability of large systems. BioSystems 17: 171–177

Hayden B (1981) Research and development in the Stone Age: technological transitions among hunter-gatherers. Curr Anthropol 22: 519–548

Hempel CG (1965) Aspects of scientific explanation. Free Press, NY

Hinde RA (1982) Ethology: its nature and relations with other sciences. Oxford Univ Press, Oxford

Hirsch M, Smale S (1974) Differential equations, dynamical systems, and linear algebra. Academic Press, Lond NY

Hofbauer J (1981) On the occurrence of limit cycles in the Lotka-Volterra equation. Nonlinear Anal 5: 1003–1007

Hofbauer J, Sigmund K (1984) Evolutionstheorie und dynamische Systeme. Parey, Berlin (transl) (1988) The theory of evolution and dynamical systems, Cambridge Univ Press, Cambridge

Hofbauer J, Schuster P, Sigmund K (1979) A note on evolutionarily stable strategies and game dynamics. J Theory Biol 81: 609–612

Hutchinson GE (1947) A note on the theory of competition between two social species. Ecology 28: 319–321

Huxley J (1942) Evolution: the modern synthesis. Allen and Unwin, Lond

Irons W (1979) Natural selection, adaptation, and human social behavior. In: Chagnon NA, Irons W (eds) Evolutionary biology and human social behavior. Duxbury Press, North Scituate, Mass, pp. 4–39

Johnson GR (1986) Kin selection, socialization, and patriotism: an integrating theory. Politics life sci 4: 127–140

Jones DS (1979) Elementary information theory. Clarendon, Oxford

Kanitscheider B (Hrsg) (1979) Materie – Leben – Geist. Zum Problem der Reduktion der Wissenschaften. Duncker und Humblot Ber

Kitcher P (1985) Vaulting ambition: sociobiology and the quest for human nature. MIT Press, Cambridge, Mass

Krebs JR, Davies NB (eds) (1978) Behavioural ecology. Blackwell, Oxford

Krebs JR Davies, NB (1981) An introduction to behavioural ecology. Blackwell, Oxford

Küppers B-O (1983) Molecular theory of evolution. Springer, Berlin Heidelberg New York

Küppers B-O (1984) Das 'Paradoxon' der Evolution. In: Kanitscheider B (Hrsg) Moderne Naturphilosophie. Königshausen und Neumann, Würzburg, 317–335

LaSalle J, Lefshetz S, (1961) Stability by Lyapounov's direct method. Academic Press, Lond NY

Leach E (1981) Biology or social science: wedding or rape? Nature (Lond) 291: 267–268

Lewontin RC (1979) Sociobiology as an adaptationist program. Behav Sci 24: 5–14

Lewontin RC (1983) Gene, organism and environment. In: Bendall DE (ed) Evolution from molecules to men. Cambridge Univ Press, Cambridge, pp 273–285

Lewontin RC (1983) Keeping it clean (book review). Nature (Lond) 300: 113–114

Lex BW (1978) Neurological bases of revitalization movements. Zygon 13: 276–312

Lex BW (1979) The neurobiology of ritual trance. In: d'Aquili EG, Laughlin CD, McManus J (eds) The spectrum of ritual. Columbia Univ Press, NY, pp. 117–151

Lorenz K (1973) Die Rückseite des Spiegels. Piper, München. (transl) (1977) Behind the mirror. Methuen, Lond

Losco J (1981) Ultimate vs proximate explanation. J Social Biol Struct 4: 329–346

Lumsden CJ, Wilson EO (1981) Genes, mind, and culture. Harvard Univ Press, Cambridge, Mass

Martin P, Bateson P (1986) Measuring behavior. Cambridge Univ Press, Cambridge

Masters RD (1981) The value – and limits – of sociobiology. In: White E (ed) Sociobiology and human politics. Heath, Lexington, Mass, pp. 135–165

Masters RD (1983) The biological nature of the state. World Politics 35: 161–193

Maull N (1977) Unifying science without reduction. Stud Hist Phil Sci 9: 143–162

May RM (1975) Biological populations obeying difference equations: stable points, stable cycles, and chaos. J Theor Biol 51: 511–524

May RM (1976) Simple mathematical models with very complicated dynamics. Nature (Lond) 261: 459–467

Maynard Smith J (1964) Group selection and kin selection. Nature (Lond) 201: 1145–1147

Maynard Smith J (1976) Group selection. Q Rev Biol 51: 277–283
Maynard Smith J (1982) Evolution and the theory of games. Cambridge Univ Press, Cambridge
Maynard Smith J, Parker GA (1976) The logic of asymmetric contests. Anim Behav 24: 159–175
Maynard Smith J, Price GR (1973) The logic of animal conflict. Nature (Lond) 246: 15–18
Mayo O (1983) Natural selection and its constraints. Academic Press, Lond NY
Mayr E (1954) Change of genetic environment and evolution. In: Huxley J (ed) Evolution as a process. Allen and Unwin, Lond, pp. 351–376
McCleery RH (1978) Optimal behaviour sequences and decision making. In: Krebs JR, Davies NB (eds) Behavioural ecology: an evolutionary approach. Blackwell, Oxford, pp 377–410
McEachron DL, Baer D (1982) A review of selected sociobiological principles: application to hominid evolution. II. The effects of intergroup conflict. J Social Biol Struct 5: 121–139
Mesarović MD, Takahara Y (1975) General systems theory: mathematical foundations. Academic Press, Lond NY
Mesarović MD, Macko D, Takahara Y (1970) Theory of hierarchical, multilevel, systems. Academic Press, Lond NY
Milkman R (1982) Toward a unified selection theory. In: Milkman R (ed) Perspectives on evolution. Sinauer, Sunderland, Mass, pp. 105–118
Monod J (1970) Le hasard et la nécessite'. Édition du Seuil, Paris
Montagu A (ed) (1980) Sociobiology examined. Oxford Univ Press, Oxford
Nagel E (1979) The structure of science, 2nd edn. Hackett, Indianapolis
Nagylaki T, Crow J (1974) Continuous selective models. Theor Popul Biol 5: 257–283
Nei M (1987) Molecular evolutionary genetics. Columbia Univ Press, NY
Netting RM (1971) The ecological approach in cultural study. Addison-Wesley, Reading, Mass
Nicolis G, Prigogine I (1977) Self-organization and non-equilibrium systems. Wiley, NY
Odum EP (1969) The strategy of ecosystem development. Science 164: 262–270
Olsen LF, Degn H (1985) Chaos in biological systems. Q Rev Biophys 18: 165–225
Omark DR, Strayer FF, Freedman DG (eds) (1980) Dominance relations. Garland, NY
O'Neill RV, DeAngelis DL, Waide JB, Allen TFH (eds) (1986) A hierarchical concept of ecosystems. Princeton Univ Press, Princeton
Oster GF, Wilson EO (1978) Caste and ecology in the social insects. Princeton Univ Press, Princeton
Oster GF, Ipaktchi A, Rocklin S (1976) Phenotypic structure and bifurcation behavior of population models. Theor Popul Biol 10: 365–382
Parker GA (1974) Assessment strategy and the evolution of animal conflict. J. Theory Biol 47: 223–243
Partridge L (1978) Habitat selection. In: Krebs JR, Davies NB (eds) Behavioural ecology: an evolutionary approach. Blackwell, Oxford, pp 351–376
Pearce D (1982) The structuralist concept of reduction. Erkenntnis 18: 307–333
Pennock JR, Chapman JW (eds) (1972) Coercion. Lieber-Atherton, NY
Phillips DC (1976) Holistic thought in social science. Stanford Univ Press, Stanford
Pichler F (1975) Mathematische Systemtheorie. De Gruyter, Berl
Pilbeam D (1972) The ascent of man. Macmillan, NY
Popp J, Devore L (1979) Aggressive competition and social dominance theory: synopsis. In: Hamburg D, McCown E (eds) The great apes. Benjamin-Cummings, Menlo Park, Ca, pp. 317–338
Popper K (1977) The self and its brain I. In: Popper K, Eccles JC. The self and its brain. Springer, Berlin Heidelberg New York, pp 3–223
Popper K, Eccles JC (1977) The self and its brain. Springer, Berlin Heidelberg New York
Poston T, Stewart I (1978) Catastrophe theory and its applications. Pitman, Lond
Prigogine I (1980) From being to becoming: time and complexity in physical sciences. Freeman, San Francisco
Prigogine I, Stengers I (1980) Dialogue with nature. Doubleday, NY
Rapoport A (1986) General system theory. Abacus, Turnbridge Wells
Ridley M (1983) The explanation of organic diversity. Clarendon, Oxford
Riedl R (1979) Biologie der Erkenntnis. Parey, Berl
Rose S, Lewontin RC, Kamin LJ (1984) Note in our genes. Penguin Books, Harmondsworth
Rosen R (1978) Fundamentals of measurement and representation in natural systems. Elsevier – North Holland, NY

Rosie AM (1973) Information and communication theory. Van Nostrand Reinhold, Lond
Roughgarden JH (1979) Theory of population genetics and evolutionary ecology. Macmillan, NY
Sahlins M (1976) The use and abuse of biology. Univ Michigan Press, Ann Arbor
Sattler R (1986) Biophilosophy: analytic and holistic perspectives. Springer, Berlin Heidelberg New York Tokyo
Schaffner KF (1967) Approaches to reduction. Phil Sci 34: 137–147
Schmidt J (1973) Mengenlehre. Bibliogr Inst, Mannheim
Schubert G (1983) The structure of attention: a critical review. J Social Biol Struct 6: 65–80
Schuster P, Sigmund K (1983) Replicator dynamics. J Theory Biol 100: 533–538
Schuster P, Sigmund K, Hofbauer J, Wolff R, Gottlieb R, Merz P (1981) Self-regulation of behaviour in animal societies I–III. Biol Cybern 40: 1–25
Schwabhäuser W (1971) Modelltheorie I. Bibliogr Inst, Mannheim
Schwabhäuser W (1972) Modelltheorie II. Bibliogr Inst, Mannheim
Serra R, Andretta M, Companiani M, Zanarini G (1986) Introduction to the physics of complex systems. Pergamon, Oxford
Service ER (1975) Origins of the state and civilization. Norton, NY
Simpson GG (1944) Tempo and mode in evolution. Columbia Univ Press, NY
Slatkin, M (1979) The evolutionary response to frequency- and density-dependent interactions. Am Nat 114: 384–398
Smale S (1976) On the differential equations of species in competition. J Math Biol 3: 5–7
Sober E (1984a) The nature of selection. MIT Press, Cambridge, Mass
Sober E (ed) (1984b) Conceptual issues in evolutionary biology, MIT Press, Cambridge, Mass
Sociobiology Study Group (1977) Sociobiology – a new biological determinism. In: The Ann Arbor Sci People Editorial Collect (ed) Biology as a social weapon. Burgess, Ann Arbor, pp 133–153
Sperry RW (1965) Mind, brain, and humanist values. In: Platt JR (ed) New views on the nature of man. Chicago Univ Press, Chicago, pp 71–92
Stebbins GL (1982) Darwin to DNA, molecules to humanity. Freeman, San Francisco
Steward JH (1968) Cultural ecology. In: International Encyclopedia of the Social Sciences, Vol 4. Macmillan, NY, pp. 337–344
Stewart OC (1956) Fire as the first great force employed by man. In: Thomas WL Jr (ed) Man's role in changing the face of the earth. Chicago Univ Press, Chicago, pp 115–133
Taylor PD (1979) Evolutionarily stable strategies with two types of players. J. Appl Prob 16: 76–83
Taylor PD, Jonker LB (1978) Evolutionarily stable strategies and game dynamics. Math Biosci 40: 145–156
Templeton AR (1980a) Modes of speciation and inferences based on genetic distance. Evolution 34: 719–729
Templeton AR (1980b) The theory of speciation via the founder principle. Genetics 94: 1011–1038
Templeton AR (1982) Adaptation and the integration of evolutionary forces. In: Milkman R (ed) Perspectives on evolution. Sinauer, Sunderland, Mass, pp. 15–31
Thom R (1975) Structural stability and morphogenesis. Benjamin – Addison Wesley, NY
Thomas B (1985a) Genetical ESS-models: I Concepts and basic models. Theor Popul Biol 28: 18–32
Thomas B (1985b) Genetical Ess-models: II. Multi-strategy models and multiple alleles. Theor Popul Biol 28: 33–49
Thomas B, Pohley HJ (1982) On a global representation of dynamical characteristics in ESS-models. BioSystems 15: 141–153
Trivers R (1981) Sociobiology and politics. In: White E (ed) Sociobiology and human politics. Heath, Lexington, Mass, pp. 1–43
Trivers R (1985) Social evolution. Benjamin-Cummings, Menlo Park, Ca
Trivers R, Hare H (1976) Haplodiploidy and the evolution of the social insects. Science 191: 249–263
Uyenoyama M, Feldman MW (1980) Theories of kin and group selection: a population genetics perspective. Theor Popul Biol 17: 380–414
van den Berghe P (1981) The ethnic phenomenon. Elsevier – North Holland, NY
von Bertalanffy L (1968) General system theory. Braziller, NY
von Kutschera F (1972) Wissenschaftstheorie I, II. Fink (UTB), München
Weber M (1958) Wirtschaftsgeschichte. Abriß der universalen Sozial- und Wirtschaftsgeschichte, 3. Aufl. Mohr, Tübingen
Weber M (1972) Wirtschaft und Gesellschaft, 5. Aufl. Mohr, Tübingen

Weber M (1977) Politik als Beruf, 6. Aufl. Duncker und Humblot Berl
Webster D (1975) Warfare and the evolution of the state: a reconsideration. Am Antiqu 40: 464–470
Weidlich W, Haag G (1983) Concepts and models of a quantitative sociology. Springer, Berlin Heidelberg New York
West MJ (1967) Foundress associations in polestine wasps. Science 157: 1584–1585
West Eberhard MJ (1975) The evolution of social behavior by kin selection. Q Rev Biol 50: 1–34
White E (ed) (1981) Sociobiology and human politics. Heath, Lexington, Mass
White LA (1954) The energy theory of cultural development. In: Kapadia KM (ed) Professor Ghurye Felicitation Volume. Popular Book Depot, Bombay pp 1–8
White LA (1959) The evolution of culture. McGraw-Hill, NY
Willhoite FH Jr (1981) Rank and reciprocity: speculations on human emotions and human politics. In: White E (ed) Sociobiology and human politics. Heath, Lexington, Mass, pp 239–258
Williams GC (1974) Adaptation and natural selection, 1st pbk edn Princeton Univ Press, Princeton
Wilson DS (1980) The natural selection of populations and communities. Benjamin-Cummings, Menlo Park, Ca
Wilson EO (1971) The insect societies. Belknap, Cambridge, Mass
Wilson EO (1975) Sociobiology: the new synthesis. Belknap, Cambridge, Mass
Wimsatt, WC (1974) Complexity and organization. In: Schaffner KF, Cohen RS (eds) Boston studies in the philosophy of science XX. Reidel, Dordrecht, pp 67–86
Wimsatt WC (1975) Reductionism, levels of organization, and the mind-body problem. In: Globus GG, Maxwell G, Savodnik I (ed) Consciousness and the brain. Plenum, NY
Wimsatt WC (1976) Reductive explanation: a functional account. In: Cohen RS, Wartofsky MW, (eds) Boston studies in the philosophy of science XXXII, Reidel, Dordrecht, pp 671–710
Wimsatt WC (1979) Reduction and reductionism. In: Asquith PD, Kyburg HE Jr (eds) Current research in philosophy of science. Philos Sci Assoc, East Lansing, pp 352–377
Winterhalder B (1980) Environmental analysis in human evolution and adaptation research. Hum Ecol 8: 135–170
Wittenberger JF (1981) Animal social behavior. Duxbury, Boston
Woodger JH (1959) Studies in the foundations of genetics. In: Henkin L, Suppes P, Tarski A (eds) The axiomatic method. North Holland, Amst, pp. 408–428
Wright HT (1977) Recent research on the origins of the state. Ann Rev Anthropol 6: 379–397
Wright S (1980) Genic and organismic selection. Evolution 34: 825–843
Zadeh LA (1969) The concepts of system, aggregate and state in system theory. In: Zadeh LA, Polak E (eds) System theory. McGraw-Hill, NY, pp 3–43
Zeeman EC (1980) Population dynamics from game theory. In. Nitecki Z, Robinson C (eds) Global theory of dynamical systems. Springer, Berlin Heidelberg New York, pp 471–497
Zeeman EC (1981) Dynamics of the evolution of animal conflict. J Theory Biol 89: 249–270

List of Symbols

Logic:

¬	Negation
∧	conjunction
∨	disjunction
⇒	material implication
⇔	material equivalence
⋀	universal quantifier
⋁	existential quantifier
$[\![(.,\ldots,.)]\!]$	list of free variables

Set Theory:

∈	element relationship
{ }	set
∅	empty set
∩	intersection
∪	union
⊂	inclusion (including identity!)
×	Cartesian product
M^n	n-fold Cartesian product of set M
$\mathscr{P}$	power set
card	cardinality
$[A_i]_{i \in I}$	family of objects, sets, etc., A_i with index set I
→	mapping, function
$\mathbb{P}$	set of positive integers
$\mathbb{R}$	set of real numbers
$\mathbb{R}_0^+$	set of non-negative real numbers
$\langle .,\ldots,. \rangle$	vector, finite sequence
$\rangle\ \langle$	function assigning the m-tuple $\rangle z\langle = \langle \{x_1\},\ldots,\{x_m\} \rangle$ to the m-tuple $z = \langle x_1,\ldots,x_m \rangle$
$[\![\ ]\!]$	function assigning $[\![A]\!] = B_1 \times \cdots \times B_m$ to the m-tuple $A = \langle B_1,\ldots,B_m \rangle$ of sets

Subscript dots indicate expectation values (statistical averages) of vectors and tensors.

Subject Index

Adaptation   7, 59-60, 73-78, 102, 109, 112, 115-116, 120, 123, 131, 137-138, 146-149, 152
Adaptive evolution   7, 37, 59, 73-78, 83-86, 131-134, 138
Adaptive topography   114-116
Altruism   74, 77, 93-97, 132, 148
Analogous adaptations   74, 85-86, 93, 95, 102, 139, 151
Asymmetric games   86, 100, 141-143, 145
Asymptotic stability   79-80, 82, 84, 86, 89-90, 98-101
Attractor *see* Asymptotic stability
Authority   141-142, 144-146, 148, 150
Autocatalysis   26, 117-119, 152
Axiom   42-46, 48, 54

Bifurcation *see* Structural stability (instability)
Biocultural coevolution   117-118, 122-130, 133

Catastrophe (mathematical)   103, 107, 111, 114-115
Coercive power   140, 142-143, 145-146, 151
Coevolution   73, 117, 119, 127, 133
Competition   67-68, 87, 95, 103, 110-112, 116, 118, 121, 123, 127, 129, 149
Competitive exclusion   109, 111, 125-126
Complexity   7-8, 20, 23, 25-26, 57-58, 69, 75, 79, 117, 133-136, 139, 147-148, 151
  behavioural   22, 37, 57
  commensurability   22, 25, 38, 46, 51
  comparative concept of   20-21, 23, 25, 36-38, 57
  structural   22, 36-37, 58, 104
Connexion of systems *see* Coupling relation
Control parameter *see* State parameter
Co-operation   73, 87, 96, 132-134, 141-142, 146, 149
Coupling degree   19-21, 29, 56-57
  relation   14-17, 19-21, 24, 29, 67, 69
Cultural capacity *see* Culture
  ecology   118, 122, 152
Culture   1, 60, 63, 117-118, 120, 122-134, 138-139, 146-148, 151

Dominance, social   132, 141-143, 148-149
Domination   140, 142-146, 150
Dynamic stability   79-80

Ecological niche   108-109, 111, 116, 118
  release   119, 125
Effect
  causal   32-33, 36, 51, 59, 63
  deterministic   32-33, 35-36, 59
Emergence   26-27, 29, 32-40, 56, 59, 62, 152
Entropy   13, 20, 23, 77
Equilibrium   66, 81-84, 86, 89-91, 98-101, 107, 113-115, 119-120, 128-129
Ethology   2, 63, 73-74, 141, 149, 151
Evolution (*see also* Self-organisation)   1-2, 4, 7, 22, 25-29, 35-39, 56-58, 69, 147-148
  biosocial   4, 73-78, 83, 131, 150, 152
  organic   7, 36, 101-102, 116, 118
  secular   78, 97-104, 107-108, 112, 115-116, 120, 123, 152
  sociocultural   36, 117, 133-134, 143, 148, 150-152
Evolutionarily stable strategy (ESS)   81-85, 97-102, 137, 139, 141
Evolutionary discontinuity   102-104, 107, 112, 114-115, 117
  game theory   4, 35, 74-75, 80-86, 97, 137, 139, 141, 143
  gradualism   102-104, 106-107, 109, 112, 115
  stability (instability)   4, 78-83, 97-102, 131-139
Evolutionism   1, 26, 152
Evolved *see* Complexity

Family (set-theoretic)   10, 16, 53, 59, 69
  (social) *see* Kinship

Gene selection   75, 77, 84, 93, 95, 102
Group selection   74-75

Hamilton's rule   93-97
Hierarchical differentiation *see* Evolution
History of civilisation   128-129
Holism   2-4, 26-27, 29, 32, 34-36, 38, 40,

Holism 59-60, 133, 152
Homologous traits 74, 151

Inclusive fitness 75-76, 85, 90, 93-96, 102, 132-133, 138, 146, 149
Individual selection 74-75
Influence 140-142
Input-output system 2, 13, 16, 20, 24, 36, 52-54

Kin selection 75, 93-97
Kinship 132-133, 143, 145-146, 149

Language, formal 30, 41, 42-50, 63
Law (juridical) 145-146, 150-151
  (scientific) 32-33, 42-45, 52, 56, 152
Learning 118-119, 122, 125-131, 142
Lyapounov function 99, 107, 110, 113-114

Macrostructure 1-3, 33, 39, 51, 138
Macrosystem 18, 25, 135
Mendelian inheritance 41, 63-64
Microstructure 1-3, 33, 51, 138
Microsystem 18, 25, 135
Model, semantic 30, 45-46
  theory (see also Semantic interpretation) 2, 45-50, 58
Mutation 84-86, 89-101, 108, 112, 118-119, 127

Neo-Darwinian evolution 7, 73-78, 103
  theory 41, 63-64, 77-78, 84, 102, 121, 131, 133-134, 152

Obscurantism VII, 40
Organisation see Structure
Organisational differentiation see Evolution
Organismic selection see Individual selection

Parameter family of systems 41, 52, 58, 65-69
Parametrisation of laws 2, 42-46, 58, 152
  of systems 52-56, 65-69, 79, 83, 152
Phase 66-67, 79
Phase portrait 67, 104
Phenotype profile 105-107
Political organisation 134-135, 140, 143, 146, 148, 150
Power 138, 140-142
  political 133-135, 139-140, 143-148, 151
Preadaptation 137, 143, 148-150
Process 25, 57
  evolutionary 7, 10, 22, 25, 28, 37, 39, 56-58, 65, 69, 152
Promotion 147-150
Punctuated equilibria see Evolutionary discontinuity
Punctuational change see Evolutionary discontinuity

Rank society 134, 141, 143, 145, 149, 151
Reduction of theories 3, 29-31, 34-36, 40-42, 59
  simultaneous 40-42
Reductionism 1, 3-4, 27, 29, 34, 36, 38, 41, 59, 75-76, 133
Relation 10-11, 20
  homogeneous 11, 19-20, 27
  inhomogeneous 11
  input 13-14, 16, 24-25, 52
  output 13-14, 16, 24-25, 52
Representation 32-34, 51
  lower-level 32-33, 38-39, 45, 51
  macroscopic 3, 18
  microscopic 3, 18, 38-39
Representation Theorem 3, 45-51, 56
Reproduction, organismic 88, 93, 95-97, 149
  rate of 104, 118-120, 122
Ritual 63, 132, 141-142, 145, 149-151
Ritualisation 63, 73-75, 141, 148, 150

Selection 7, 40-41, 64, 73-78, 83-86, 89, 93, 95, 97-103, 115-116, 121-122, 129, 131, 137, 146, 148, 152
Self-organisation, hierarchical 1, 3, 20, 22, 134-136
  natural 3, 20, 28-29, 36, 56, 77, 134, 136, 152
  sociocultural 118, 134
Semantic interpretation 12, 30-31, 34, 37, 40-50, 54-56, 63
Shifting peak 103, 106, 110
Similarity 58-61, 112, 117
Simulation of systems 60-63
Social behaviour 73-78, 82, 93-97, 101-102, 131-135, 141
Society 1, 138
  complex 134, 139, 143-144, 147-151
  primitive 134, 139, 143, 148-150
Sociobiology 4, 26, 36, 40, 60, 73-78, 83-85, 95, 101-102, 131-134, 139, 147, 149, 151-152
State (political) 134, 139, 143-150
  (system-theoretic) 13-14, 43, 53-58, 65
State-determined 55-58, 65-69
State parameter 14, 43, 104, 107, 114-115, 123, 152
Stationarity see Equilibrium
Sterility 88, 93-96
Stratum 13, 19-20, 27, 32-35, 38-39, 46, 51, 56, 59
Structural stability (instability) 51, 78-86,

91-93, 97, 100-102, 104, 106-108, 110, 112-114, 131-139, 143, 145, 152-153
Structure 7, 11-12, 24, 36, 52-53
  hierarchical 12-13, 20, 22-23, 25-29, 32, 34-36, 38-39, 41, 51, 56, 133-136, 148
  macroscopic *see* macrostructure
  microscopic *see* microstructure
  semantic 2, 12, 34-35, 52-54, 58, 76
  social 131-135, 137-140, 142, 148-149
Subsystem 17, 24, 57
  coupled 16, 56-57, 60, 68
  decoupled 17, 21, 27, 56-57, 60, 68
Symmetric games 80, 86, 141
Synthesis of theories 30, 40-45, 58, 64
System 13-14, 24
  component 13, 16, 22, 57, 68, 138
  coupled 14-16, 19-21, 27, 56, 60, 67
  deterministic 3, 51

  dynamical 3, 51, 58, 65-68, 79-86, 136-137, 143, 148, 152-153
  hierarchical 3, 17-18, 20-21, 25, 28, 32, 34, 45, 51-52, 55-58
  stochastic 3, 148, 152
  stratified *see* hierarchical
  theory 2, 10, 26, 52, 58, 152

Theory 2, 11, 29-31, 36-37, 40-51, 63
  complementary 48
  deterministic 3, 29, 34, 51
  elementary 3, 29-30, 34, 37, 41, 52, 58, 76
  first-order *see* elementary
  stochastic 38-39, 51, 136, 148
  unified 2, 40-42, 44, 46, 54-58, 83, 131-132, 152
Tradition *see* Learning